PRAISE FOR *PACIFIC NORTHWEST CHEESE: A HISTORY*

"Tami Parr's research reveals the roots of today's artisan cheese renaissance: the people, the geography, and the historical forces that shaped cheese production in the Pacific Northwest. An impressive and valuable work of scholarship."

—Janet Fletcher, author of *Cheese & Beer* and *Cheese & Wine*

"Over many years, Tami Parr has made significant contributions to the success of American artisan cheese. Her new book clarifies the history of cheesemaking in the Pacific Northwest, explains the critical value of cheese as a basic food for early settlers, and documents vibrant artisan production in the region throughout the 20th century. Equally important, her analysis contributes to a deeper understanding of American cheese history."

—Jeff Roberts, author of *Atlas of American Artisan Cheese*

"As a farmstead cheese maker in the Pacific Northwest, I must admit that I rather mistakenly thought that cheese was invented right around the time that I started milking cows and making cheese. *Pacific Northwest Cheese: A History* has shown me how mistaken I was and has given me a new perspective on the origins of cheese in this region. I was especially intrigued by the recent history—that of the 1970s back-to-the-land movement. It is reassuring to see that many have come before me: striking out on a few acres with a small herd of cows and learning to produce fine cheeses."

—Kurt Timmermeister, author of *Growing a Farmer: How I Learned to Live off the Land*

D1534439

Pacific Northwest Cheese: A History

TAMI PARR

Oregon State University Press ✐ Corvallis

The paper in this book meets the guidelines for permanence and durability of the Committee on Production Guidelines for Book Longevity of the Council on Library Resources and the minimum requirements of the American National Standard for Permanence of Paper for Printed Library Materials Z39.48-1984.

Library of Congress Cataloging-in-Publication Data

Parr, Tami J.
Pacific Northwest cheese: a history / Tami Parr.
    pages cm
  Includes bibliographical references and index.
  ISBN 978-0-87071-704-8 (alk. paper) -- ISBN 978-0-87071-705-5 (ebook)
  1. Cheese--Northwest, Pacific--History. 2. Cheesemakers--Northwest,
Pacific--History. 3. Cheese factories--Northwest, Pacific--History. I. Title.
SF274.U6P373 2013
637'.309795--dc23
                              2013004369

Oregon State University Press
121 The Valley Library
Corvallis OR 97331-4501
541-737-3166 • fax 541-737-3170
www.osupress.oregonstate.edu

Contents

Acknowledgments 7

Introduction 9

1 Furs, Cattle, and Empire: English Cheese in the Pacific Northwest . . . 13

2 Milk and Cheese in Oregon Country 35

3 From Farm to Factory: Cheese Becomes Big Business 55

4 Expansion and Innovation 83

5 The Mass Production Era 109

6 Back to the Farm: The Artisan Cheese Renaissance 133

   APPENDIXES

A A Short History of Cheese in Alaska 163

B Artisan Cheesemakers of the Pacific Northwest 2012 169

   Notes and Sources 173

   Index 197

Acknowledgments

In the course of writing this book I worked with many dedicated people at county, state, and regional historical societies across the Pacific Northwest. I can't say enough about how valuable these institutions are to preserving the history of communities across the region. Many thanks to all of the researchers and volunteers who took the time to help me in uncovering information and photos that document regional dairy and cheese history. Thanks also to the librarians and staff at the special collections departments I visited who provided assistance and advice. These knowledgeable men and women are also an incredibly valuable resource.

A number of others have provided invaluable information, advice, and assistance as I wrote this book. Many thanks to Chandra Allen of Tillamook County Creamery Association, Marc Bates, Floyd Bodyfelt, Gianaclis Caldwell, Laurie Carlson, the Castrilli family, Sasha Davies, cartographer Grace Gardner, Earl Gilmartin, Lisbeth Goddik, David Gremmels, Angie Jabine, Jeff Kronenberg, Bill Moomau, Bill Predeek, Jeff Roberts, Russ Salvadalena, John Shuman, and Theresa Simpson.

Thank you also to Mary Braun, Jo Alexander, and the generous folks at Oregon State University Press for their enthusiasm and support for this project.

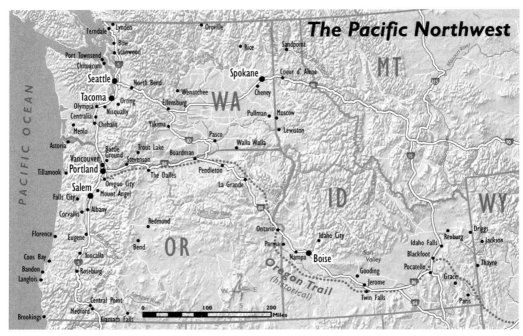

The Pacific Northwest

PACIFIC OCEAN

WA

OR

ID

MT

WY

Ferndale • Lynden • Oroville
Bow • Rice • Sandpoint
Port Townsend • Stanwood
Chimacum
Seattle • North Bend • Spokane • Coeur d' Alene
Tacoma • Orting • Wenatchee • Cheney
Olympia • Nisqually • Ellensburg • Pullman • Moscow
Centralia • Chehalis • Yakima • Lewiston
Menlo • Pasco
Astoria • Walla Walla
Battle Trout Lake • Boardman
Vancouver • Ground • Stevenson
Tillamook • Portland • The Dalles • Pendleton
Salem • Oregon City • La Grande
Falls City • Mount Angel
Corvallis • Albany • Redmond • Ontario • Idaho City
Florence • Parma • Idaho Falls • Rexburg • Driggs
Eugene • Bend • Nampa • Boise • Blackfoot • Jackson
Coos Bay • Yoncalla • Sun • Pocatello • Thayne
Bandon • Roseburg • Valley • Gooding • Grace
Langlois • Jerome • Paris
Central Point • Twin Falls
Brookings • Medford
Klamath Falls

OR

John Day River
Santiam River
Willamette River
Umpqua River
Missouri River
Oregon Trail (historical)

0   100   200
Miles

Gardner Cartography (2012)

Introduction

If you've visited a farmers market lately you have probably come across more than one local cheesemaker selling their wares. Using the milk of goats, sheep, and cows, cheesemakers in Oregon, Washington, and Idaho—eighty-six of them as of this writing—are producing an incredible and delicious variety of fresh and aged cheeses.

While writing my first book, *Artisan Cheese of the Pacific Northwest*, I traveled all around Oregon, Washington, Idaho, and British Columbia visiting farms and talking to the many hardworking, dedicated people making cheese in the region. I learned a lot about cheese (and ate a lot of cheese) during that time, but I also discovered that there was more to the story of regional cheese than meets the eye. Every once in a while someone would mention a tidbit of an interesting story from the past: something about a gouda cheesemaker near Yakima, Washington; murmurs about an abandoned cheese cave near Mt. Adams in Washington; or whispers of a goat's milk cooperative that made cheese in Salem, Oregon, back in the '70s. I knew I had to find out more.

Once I started digging, I found a gold mine. The story of cheese in the Pacific Northwest goes as far back as the mid-nineteenth century, when the area was still an amorphous swath of land west of the Louisiana Purchase and north of Mexico, which then encompassed the present-day states of California, Nevada, and Utah. European powers, blithely ignoring the Native populations, vied for control over the region until the international border between the United States and Canada that we know today was established in 1846. The states of Oregon, Washington, and Idaho later emerged out of the chaos and have come to form what we call the Pacific Northwest today.

Because the Pacific Northwest was settled late relative to other parts of the United States, the region's dairy history is a bit more condensed than that of dairy and cheesemaking states farther east like Wisconsin, New York, or Vermont. Still, the story of cheese in the region covers a lot of ground. It starts out among fur traders-turned-farmers from the Hudson's Bay Company who brought over English dairy workers, some of them women, to make cheese at makeshift dairies carved out of the wilderness. This cheese, which I like to think of as "colonization cheddar," was an apt symbol of the encroachment on the area's indigenous peoples by Europeans. Decades later, settlers who dragged themselves and their livestock to the Pacific Northwest over the Oregon Trail kept cows and made and sold butter and cheese as a means of survival. Soon enough, industrialization led to an era of mass production and cheese became a profitable commodity that was produced in enormous quantities. Successive waves of entrepreneurs, the most famous of them J. L. Kraft, saw the latent dairying potential in the wide open spaces of the Pacific Northwest and wasted no time in taking advantage of it.

While by the mid-twentieth century factory-made cheese was the norm across the United States, during the 1960s and '70s that state of affairs slowly began to change. Renewed interest in land stewardship and wholesome, natural food led to, among other things, a rebirth of small-scale farmstead cheesemaking. This modern wave of cheesemaking has continued to expand into the phenomenon we see today: a full-fledged artisan cheese renaissance. Today hundreds of cheesemakers across the country are busy every day resurrecting the ancient craft of turning milk into delicious cheese. We are lucky to have so many of them here in the Pacific Northwest.

Though most traces of the region's early cheese and dairy industry have long since disappeared, remnants are everywhere. On the northern Oregon coast, a number of old cheese factories and creamery buildings still stand, relics of the days when cheese used to be made under the auspices of the Tillamook cooperative. In Pocatello, Idaho, the old Kraft factory, which opened in 1924 and produced processed cheese for decades, now sits abandoned. In the case of Rogue Creamery in Central Point in southern Oregon, the old cheese factory is not only still there, it's still making cheese. These and other buildings like them are relics of a past era when the local creamery or cheese factory was the hub of the economy in many rural towns.

This project grew out of a love of cheese and a curiosity about the people, places, and events that have shaped the regional cheese industry as it exists today. Ultimately, I hope to have shed some light on the intriguing history of regional cheesemaking as it has evolved over nearly two hundred years. I also hope this book inspires readers to find out more about the creameries and cheese factories that may have once been located in the far corners of the Pacific Northwest. Lastly, I'd also like to encourage readers to get out and enjoy some locally produced artisan cheese right now, because that's cheese history happening right before your eyes.

Craig Nelson salts fresh cheese curds at Rogue Creamery, Central Point, Oregon, 2008. Photo by the author.

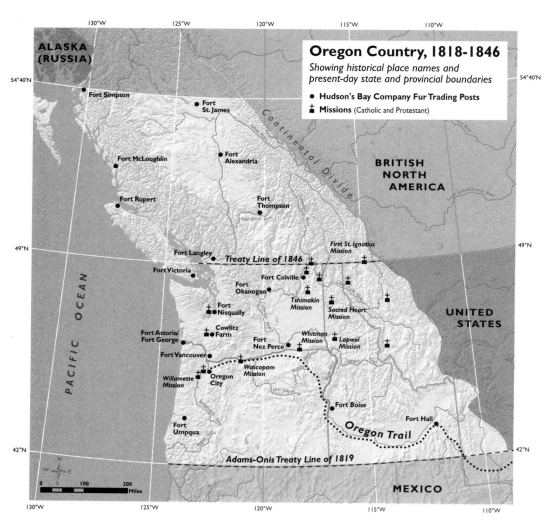

Oregon Country, 1818-1846

*Showing historical place names and
present-day state and provincial boundaries*

● **Hudson's Bay Company Fur Trading Posts**

‡ **Missions** (Catholic and Protestant)

ALASKA
(RUSSIA)

54°40'N

BRITISH
NORTH
AMERICA

Continental Divide

49°N

UNITED
STATES

Treaty Line of 1846

First St. Ignatius
Mission

PACIFIC OCEAN

Fort Simpson

Fort
St. James

Fort McLoughlin

Fort
Alexandria

Fort Rupert

Fort
Thompson

Fort Langley

Fort Victoria

Fort Colville

Fort
Okanogan

Fort
Nisqually

Tshimakin
Mission

Sacred Heart
Mission

Cowlitz
Farm

Fort Astoria/
Fort George

Fort Vancouver

Fort
Nez Perce

Whitman
Mission

Lapwai
Mission

Willamette
Mission

Oregon
City

Wascopam
Mission

Fort Boise

Fort
Umpqua

Oregon Trail

Fort Hall

42°N

Adams-Onis Treaty Line of 1819

MEXICO

0   100   200
Miles

Gardner Cartography (2012)

Chapter 1

Furs, Cattle, and Empire: English Cheese in the Pacific Northwest

Beginning as early as the fifteenth century a succession of European explorers sailed up and down the west coast of North America, searching for a variety of things including the elusive Northwest Passage, a sea route connecting Europe and Asia. No less a personality than famed English navigator Sir Francis Drake coursed along its shores, some argue as far north as present-day Oregon, in the *Golden Hind* in 1578. During the succeeding centuries, literally hundreds of European and Russian voyagers including Captains James Cook, John Meares, Bruno Heceta, and George Vancouver maneuvered ships along present-day Oregon, Washington, British Columbia, and Alaska, mapping numerous now-familiar points of geography and claiming them in the names of their respective countries.

One of the less-discussed aspects of this era of exploration is that most of these far-flung oceangoing expeditions regularly carried a variety of livestock including cattle, sheep, and goats. These animals are important for our purposes because they are the most commonly raised for milk production, and in order to make cheese one must first have milk. Housing animals on board ship made a certain amount of practical sense at the time since lack of food on the high seas was a potentially life-threatening issue and living, breathing animals represented a relatively sustainable food supply. Goats in particular were common cargo, convenient because of their small size and because they served as a source of both meat and fresh milk. Eighteenth-century English explorer Captain James Cook famously carried a well-traveled milk goat with him on his ship *Endeavour* that had already been around the world once with Cook's colleague Captain Samuel Wallis and his crew on the *Dolphin*. Likewise, Nancy the goat accompanied American Captain Robert Gray on several trips, including the one during which he found and named the Columbia River. Her death at sea in 1791 just prior to reaching the Columbia was significant enough to be noted in a crewman's journal: "Between the hours of 3 and 4 pm, departed this life our dear friend Nancy the Goat having been the Captain's companion on a former voyage around the Globe . . . She was lamented by those who got a share of her Milk!!" (*emphasis in the original*). While for the most part oceangoing animals remained on board, explorers sometimes dropped off goats or sheep at convenient grassy spots during their travels, leaving the animals to survive and forage on their own. Theoretically, the same ship or perhaps another from the same nation could return to the spot in the future and replenish their food supplies. The feral Arapawa goats of New Zealand are considered to be descendants of goats originally released by Captain Cook on his voyages; Hawaii's Big Island still suffers the effects of destructive feral goats, pigs, and sheep, descendants of those originally introduced by Cook and his contemporaries.

International interest in the Pacific coast of North America surged during the eighteenth century as Europeans competed for access to and control over its rich store of natural resources. By the late 1700s Spain, then a dominant world power, moved to stake a claim, motivated in part by the Russians, who were already actively exploring along the Bering Strait near present-day Alaska. The Spanish vessel *Santiago* anchored in Nootka Sound on the west side of Vancouver Island in 1774; not to be outdone, British ships followed suit and Nootka developed into a port of call and fur-trading center for a variety of nations that was active for several decades. Nootka also became home to the region's first resident cattle, goats, and sheep, which arrived on Spanish ships in 1789. Historians believe that some of those livestock were later transported to the short-lived Spanish encampment at Nuñez Gaona, founded in 1792 and abandoned soon after at what is now called Neah Bay, on the northwestern tip of the Olympic Peninsula—the first European settlement in what would later become the state of Washington.

Since cows, sheep, and goats are not native species to North America and the indigenous peoples of the area did not raise domesticated animals for food, the introduction of these animals was a potent harbinger of European expansionist aspirations in the region. According to the international custom of the period, the first nation to "discover" an area and definitively establish a settlement could claim sovereignty over it; among the accepted indicia of settlement were the construction of buildings and the presence of livestock. Not surprisingly, conflict eventually erupted among the various nations with a presence in the area. The Spanish and the British struggled over which had the most definitive territorial claim to Nootka Sound; the squabbling, which escalated into what came to be called the Nootka Crisis, was resolved, albeit tenuously, with a series of formal agreements that in essence granted both countries the right to explore and settle the area. This was but one of numerous power struggles among European nations in the region that would not be finally resolved until the mid-nineteenth century.

Of course discovery is a relative term, as what we now call the Pacific Northwest was already populated by indigenous peoples who had lived there for centuries. Nonetheless, expansion-minded Europeans perceived the area as theirs for the taking. The attitude was not entirely surprising, considering that Europeans were replaying a drama already enacted in the eastern half of North America, where they had disabled, destroyed, and displaced thousands of Native peoples in the name of civilization and progress. And European encroachers ended up taking much more than just land; historians estimate that, by the 1860s, as much as 80 percent of the region's Native population had been wiped out by diseases such as smallpox and malaria that Europeans brought with them.

The business of buying and selling animal pelts, once a thriving global enterprise, can be traced back as far as the Middle Ages. Fur-trapping and trading activity began on the east coast of North America in the late 1500s and migrated progressively farther west during the succeeding centuries thanks to an abundance of beavers, otters, and other fur-bearing mammals combined with a continuing and seemingly bottomless demand for their fur in Europe and Asia.

As European fur-trading activity grew increasingly active on the western side of the continent, Americans began to take notice. In 1810, Bostonian Nathan Winship, from a prominent merchant family with trade ties to the Russian-American Company (which had developed a significant trading presence in what we now know as Alaska), led his vessel *Albatross* up the Columbia River to a spot he named Oak Point just north of present-day Clatskanie, Oregon. There, he and his men offloaded a number of pigs and goats and started constructing a log structure in the hopes of starting a base for trade with Russian fur trappers operating in the Northern Pacific. Winship's fledgling settlement was short lived, however, as the river flooded his selected spot almost immediately and native Chinook peoples quickly ran the foreign interlopers out of the area. The earliest American venture to successfully establish a sustained physical presence in the Pacific Northwest was the John Jacob Astor-backed Pacific Fur Company. Astor, based in New York, had already made his fortune in the fur trade when he started the company in 1810. Astor's emissaries established the eponymous Fort Astoria near the mouth of the Columbia River in 1811. These first residents of present-day Astoria, Oregon, brought goats, pigs, and sheep with them on Astor's ship *Tonquin*.

Like the Spanish livestock first introduced at Nootka Sound, the animals of Fort Astoria served as both a source of food for the would-be fur traders and emblems of American settlement that would go a long way toward influencing the British to focus their own colonization efforts northward. But the more practical aspects of colonization proved a challenge to Astor's men. They found that managing livestock, constructing a fort and outbuildings, as well as establishing trading operations proved difficult; as a result, they left the animals to forage for food in the dense brush. Not surprisingly, several pigs were attacked and eaten by bears and others wandered off entirely, eventually producing a dangerous roaming pack of feral pigs. The men did manage, however, to establish a makeshift garden that produced turnips, radishes and potatoes from seed stock that had survived the ocean voyage.

For a variety of reasons, not the least of which were geographic isolation and lack of adequate supplies, the Pacific Fur Company's traders struggled to establish themselves as a commercially viable fur-trading operation. Astor sold his company to the Canadian-owned North West Company after the War of 1812 and, by late 1813, the outpost at Fort Astoria had been relinquished and was renamed Fort George. During its tenure at the mouth of the Columbia River the North West Company expanded the nascent trading center significantly, supplementing the existing stock of animals by importing two bulls and two cows from San Francisco. Alexander Henry, a North West Company partner, paid a visit to Fort George in 1814 and noted among other things the presence of a resident goat that regularly gave a pint of milk, which he found "a great luxury to our coffee[.]" Surely goat's milk was a welcome addition to the monotonous diet of salmon and sturgeon that the region's men were, for the most part, forced to endure. By 1817 the human population at Fort George had grown to one hundred fifty human inhabitants, supported by a more sizeable herd of twelve cattle along with growing numbers of pigs and goats.

Eventually most of the various European territorial claims in the Pacific Northwest were resolved. Spain and Russia both extinguished their claims by treaty in the early nineteenth century, with Russia retaining territory north of the 54th parallel, including what are now Alaska and far northern Canada, and Spain retaining what would eventually become California and the southwestern United States. This left the United States and the British as the sole remaining competitors for the vast expanse of territory in between, a state of affairs formalized in an 1818 treaty that codified an awkward sovereign limbo between the nations for several decades. Meanwhile, the underfinanced and overextended North West Company merged with the larger and more successful Hudson's Bay Company (HBC) in 1821, a move that turned HBC into the dominant fur-trading firm in the West. HBC went on to develop extensive agricultural operations out of which would emerge, among other things, the region's first locally produced cheeses.

From Fur Traders to Farmers

So wide and deep is the Hudson's Bay Company's influence that the history of the U.S. Pacific Northwest and the history of the company's activities in the region during this era are practically one and the same. First formed in 1670 and headquartered in London, the Hudson's Bay Company (HBC) had by the nineteenth century grown to dominate the North American fur trade from its continental headquarters at York Factory on Hudson's Bay in present-day Manitoba, Canada.

As had been the case for the Pacific Fur Company and the North West Company that preceded it, one of the Hudson's Bay Company's biggest challenges in developing a profitable fur-trading business in the Pacific Northwest was the practical establishment and support of the operations. Lacking the next-day deliveries we take for granted today, this meant developing an efficient supply chain that could manage the task of delivering provisions and supplies to its employees in the remote region as well as exporting the all-important money-making furs out to their eventual destination: auction in London. The overland route from HBC's headquarters to the west coast of North America, a distance of over two thousand miles, was fraught with physical dangers, not the least of which were multiple imposing mountain ranges like the Rockies and the Cascades. The trip by sea presented its own challenges: the voyage to and from either Hudson's Bay or London around the tip of South America took as long as eight months, and once ships arrived they faced the challenge of crossing the treacherous Columbia Bar, the point at which the Columbia River empties into the Pacific Ocean. The bar, still considered one of the most dangerous marine passages in the world, swallowed up many a ship of the era attempting to enter the river.

The failure of both the Pacific Fur Company and North West Company to successfully establish permanent outposts was demonstrative of the myriad obstacles impeding commercial aspirations in the Pacific Northwest. Though the challenges were formidable, the governor of northern operations for Hudson's Bay Company, George Simpson, newly appointed and determined to prove himself, was up to the task. Simpson embarked on the first of two extended overland reconnaissance journeys in 1824-25 to assess the fledgling network of outposts in the region, eleven in all, acquired from the North West Company. While in the field, Simpson observed that one of the company's biggest problems was the wasteful habits of its employees, who consumed company supplies (which had cost the company a considerable sum to transport westward) in excessive quantities. "The good people of the . . . interior of the Columbia generally have . . . shown an extraordinary predilection for European Provisions without once looking at or considering the enormous price it costs . . . such fare we cannot afford in the present times[.]" His solution to the perceived culture of waste and excess? In addition to familiar tactics such as staff reductions and restructuring, Simpson advocated to his superiors in London that Hudson's Bay Company employees take up farming:

> It has been said that Farming is no branch of the Fur Trade but I consider that every pursuit tending to leighten the Expence of the Trade is a branch thereof and that some of our Factors and Traders on the other side are better adapted for and would be more usefully employed on this side in the peaceable safe and easy occupation of Farming than in Councilling Dealing with Indians or exploring new countries for which many of them are totally unfit.

Simpson sold his superiors on farming by framing the idea not as a pragmatic survival strategy but rather as an opportunity to make a profit by selling any excess grain, meat, and other goods. In fact, this suggestion would not have been entirely novel to HBC's directors; the precedent had already been established at York Factory. The nearby Red River Settlement, started in 1811 by HBC shareholder Lord Selkirk, eventually developed into a sizeable farming operation that produced grains, fruits and vegetables, and dairy products used by the company.

One of the first practical steps Simpson took toward implementing his agricultural aspirations was organizational; he moved a number of posts acquired from the North West Company to areas more favorable to farming. In particular, Simpson positively gushed about a site he'd found near the confluence of the Willamette and Columbia rivers, noting that it was "beautifully situated on the top of a bank about 1¼ miles from the water side commanding an extensive view of the river the surrounding country and the fine plain below which is watered by two very pretty small lakes." Though the existing Fort George, located at the mouth of the Columbia River adjacent to the Pacific Ocean, made a more logical receiving point for company ships sailing to and from London, this new site, called Fort Vancouver, farther up the Columbia had the advantage of being situated near a series of arable prairies ready to be used for grazing and cultivation. The move was also strategic. Sensing that the Americans would have a more convincing claim of sovereignty over territory to the south of the Columbia River because of the precedent established at Fort Astoria, Simpson hoped to finally cement British claims to territory northward—and the development of a network of trading posts and farms would go a long way toward realizing that goal. And so were sown the seeds that would eventually grow into the earliest organized agricultural operations, including cattle ranching and dairy farms, in the Pacific Northwest.

When John McLoughlin took up residence as the manager ("Chief Factor" in HBC terminology) in late 1824 at Fort Vancouver, future prospects for the uncharted wilderness in the midst of a rainy Pacific Northwest winter must have seemed dicey at best. Yet the advantages were evident: open prairies spaced amongst the dense forests obviated the need for labor-intensive land clearing and provided immediately useable space for grazing and organized cultivation. And cultivate they did: the spring plantings of 1825 yielded a respectable nine hundred bushels of potatoes along with nine and a half bushels of peas that, along with a few beans, McLoughlin set aside for seed to be used the next season. HBC brought over laborers from the Sandwich Islands (now Hawaii) to work the land, and wheat production increased steadily from just twelve bushels in 1826 (the second full year of operation), to over five thousand bushels by the early 1840s. The barley and corn harvests also grew to average several thousand bushels per crop. McLoughlin also supervised the planting of apple, peach, fig, and pear trees, precursors of the area's modern-day fruit industry. Fruit trees turned out to grow surprisingly well, prompting visiting missionary Jason Lee, who passed through the fort in 1834, to remark that "the quantity of fruit [on the trees] is so great that many of the branches would break if they were not prevented by props." Though contemporary estimates of its overall reach varied widely, the Fort Vancouver farm operations eventually grew to extend for at least fifteen hundred acres along the Columbia River, dwarfing Fort Astoria and Fort George's significantly less ambitious kitchen gardens.

Fort Vancouver became the headquarters and supply depot for HBC's extended network of Columbia Department trading posts, eventually numbering over thirty. But Governor Simpson did not allow the other outposts to become dependent; each was expected to cultivate some soil to the extent regional conditions allowed, and the company sent Sandwich Islanders inland to provide labor to support posts with more extensive farming operations. Among these were Fort Colville near what is now Spokane, which became a center for wheat production during its years of active farming, and Fort Langley, east of the present-day Canadian city of Vancouver. Most of the other forts in the regional network kept vegetable gardens at a minimum as well as a few chickens, goats, and/or pigs in a modest effort toward self-sufficiency.

All of what HBC termed "country produce"—the wheat, peas, barley, and other crops grown at Fort Vancouver and the other HBC outposts—served a variety of purposes. Some was consumed by the residents of the fort, with the balance shipped out as supply rations to the other regional HBC outposts, and some was provided to the endless stream of visitors that passed through Fort Vancouver during its years of operation. Fort Vancouver also supplied HBC ships trading along the West Coast and with the Sandwich Islands, as well as those traveling back and forth between the Pacific Northwest and London headquarters. Eventually the agricultural operations grew to such an extent that, as George Simpson had predicted, the company was able to turn its subsistence farming project into a for-profit operation. The biggest coup was a contract signed in 1839 with the Russian-American Company, a fur-trading concern with its center of operations located in what is now Sitka, Alaska. In the contract, HBC agreed to supply the Russians with large quantities of wheat, barley, and peas, as well as meat and butter, for a period of ten years.

Following the agreement with the Russians, HBC created the Puget Sound Agricultural Company (PSAC) to own and manage its agricultural operations as a separate business entity. Though independent on paper, PSAC was effectively a corporate subsidiary since its directors were also directors of HBC and the company purchased and sold all of PSAC's goods through Fort Vancouver. PSAC's operations were centered at Cowlitz Farm and Fort Nisqually, both of which were started by HBC in the 1830s. Cowlitz Farm was located north of Fort Vancouver on the Cowlitz River near present-day Toledo, Washington, an area replete with large open prairies perfect for organized cultivation. The operation extended over a thousand acres and was the most highly organized and productive farm of any of HBC's regional operations, producing wheat, oats, butter, and other farm produce in great quantities, as well as raising pigs, cattle, sheep, and horses. Farther north in the Puget Sound area near what is now Tacoma, Washington, Fort Nisqually was conveniently situated at the northern end of the Cowlitz Portage, a trail originally used by the Native populations that HBC appropriated for travel between the Columbia River and Puget Sound. While the marshy tidelands around Fort Nisqually proved not as fertile or productive as those surrounding Cowlitz Farm and Fort Vancouver, Nisqually developed into one of the company's main livestock grazing centers. Fort Nisqually eventually grew to serve as an economic hub for the Puget Sound region, providing food and supplies to travelers and early settlers much in the same way that Fort Vancouver did to the south.

From Wheat and Barley to Milk, Butter, and Cheese

Livestock were a critical component of Governor Simpson's agricultural and commercial objectives. Domesticated animals represented not just a source of meat and milk, but also sources of useful objects and/or income, such as wool from sheep, hides for leather, and tallow for candles and soap. Toward that end, Fort Vancouver developed sizeable herds of cattle, sheep, pigs, goats, and horses. The livestock came from a number of sources. John McLoughlin brought several dozen cattle along with him from Fort George during the initial transition of HBC operations from Fort George at the mouth of the Columbia to Fort Vancouver. Others arrived by ship from England and the Sandwich Islands, or were purchased and driven north from Spanish-held California or south from Canada. After just four years of operation, Fort Vancouver had accumulated a herd of 122 cattle and, by 1837, McLoughlin reported an inventory of 685 cattle to headquarters. Fort Vancouver's herds eventually grew so large that the company sought additional grazing areas; space was located at nearby Wapato Island (once also called Multnomah Island but known today as Sauvie Island), located just across the Columbia River from the fort, as well as opposite Wapato Island near present-day Scappoose at a farm maintained by John McLoughlin's stepson, Thomas McKay. The island location was particularly convenient because the water boundary served as a natural cattle fence. For a brief period during 1839-1840, cattle were also grazed in an area south of Scappoose on the Tualatin Plains, though HBC used this area primarily for horse pasture. At its peak in 1844, Fort Vancouver maintained a stock of about thirteen hundred cattle, one-third on Wapato Island. Fort Vancouver also kept a sizeable herd of goats, numbering as many as one hundred sixteen in 1832, that likely served as a standing supply herd, with individual goats sent to sea with HBC vessels as they came and went from the Pacific Northwest. Other goats were likely milked for personal use by fort employees, as had been the practice at Fort Astoria and Fort George.

Most of HBC's regional outposts kept cattle, sheep, pigs, and goats as well. Fort Colville's herd numbered over a hundred cattle by 1840; a substantial number of horses were kept there as well as at Fort Nez Perce in the Walla Walla area for company use on the overland supply route to and from York Factory. Fort Langley in present-day British Columbia boasted another of the more substantial herds of cattle in the district, reporting an inventory of one hundred forty-one in 1844. Other significant herds of livestock were kept at Cowlitz Farm, sometimes termed Cowlitz Grazing Farm in HBC inventory records, and at Fort Nisqually. Fort Nisqually inventory records for 1840 show just eighty cattle but eight hundred twenty sheep; the numbers increased dramatically to eight thousand sheep, eight hundred cattle, and three hundred pigs by 1846.

As the earliest fur traders at Fort Astoria/Fort George had already discovered, keeping and propagating domesticated animals in the Pacific Northwest wilderness presented significant challenges. Predators like wolves and cougars were a constant threat. Because there was no time or manpower to build fences, cattle were left to roam freely and often wandered off into the wild, never to return. Pigs in particular presented a particular challenge at Fort Vancouver. "Our stock of pigs increase slowly," reported McLoughlin in a letter to headquarters, "[as] many are poisoned by eating a root that grows in these plains." McLoughlin also recognized early on that in order to develop a significant herd of any type of animal he would need to keep them alive for breeding. Especially in the first few years of Fort Vancouver's existence, McLoughlin resisted requests by captains of HBC supply ships for fresh beef to sustain their crews on the return trip, reasoning that "[t]o kill our cattle . . . [for food] would prevent our having the means of raising a sufficient supply for our future wants." Cattle were also needed as draft animals essential to cultivating the fields and running the sawmills. As a result, for the first eleven years of the fort's existence, no cattle were killed except "a bull calf or two annually for the purpose of getting rennet."

McLoughlin would have needed rennet for only one thing—making cheese. Rennet is a general term for a collection of enzymes essential to the cheesemaking process. Chymosin, the most active of the enzymes, is instrumental in coagulating milk proteins, which those skilled in the craft process in a variety of ways to create cheese. Prior to the development of standardized rennet solutions, which became available for purchase later in the nineteenth century, the only source of rennet was the fourth stomach of a calf. As a matter of procedure calves' stomachs (typically those of male calves, as female calves were raised for milk production) were collected, dried, salted, cut into strips, and then stored for later use. When rennet was needed, cheesemakers reconstituted the dried organ material in a salt water or whey solution and used the resulting liquid during the cheesemaking process.

HBC established several formal dairies and produced both butter and cheese in significant quantities. The earliest record of a building formally referred to as a "milk house" at Fort Vancouver appears in inventory records in 1832. By 1836, when Marcus and Narcissa Whitman visited en route to establishing a mission near present-day Walla Walla, Washington, Fort Vancouver's dairy operations were established enough to merit showing off to the visitors. Narcissa Whitman noted the moment: "Visited the dairy, also, where we found butter and cheese in abundance . . . they milk between fifty and sixty cows."

HBC's dairy operation expanded quickly; the company moved large numbers of cattle to Wapato Island and McKay Farm in 1838 and established dairies in both places, and by 1841 Fort Vancouver was also operating two onsite dairies of its own. Wapato Island became the center of Fort Vancouver's dairy operations, with four dairies at its peak of operations in 1844: Logie's Dairy, Taylor's Dairy, Gilbot's Dairy, and Sauvé's Dairy. Several other HBC posts kept a few cows and had enough milk to warrant a dedicated a milk house, including Fort Okanogan and Fort Nez Perces in Eastern Washington as well as at Fort Boise and Fort Hall in what would eventually become the state of Idaho. The Puget Sound Agricultural Company also operated dairies at both Cowlitz Farm and Fort Nisqually.

Butter was relatively easy to make during this period, even prior to the invention of the mechanical cream separator later in the nineteenth century. The process simply required separating the cream from the milk, at which point the cream could be skimmed off and agitated or churned to make butter. Narcissa Whitman went into some detail about the cream skimming methods she observed at Fort Vancouver on her 1836 dairy tour:

> Their pans are of an oblong square, quite large, but shallow, flareing a little, made of wood and lined with tin. In the center is a hole and a long plug. When the cream has all arisen to the surface, [they] place the pan over a tub or pail, remove the plug and the milk will all run off leaving the cream in the pan only. I think these in a large dairy must be very convenient.

After it was made, the butter was packed into small wooden barrels called firkins for easy transport on HBC ships or overland via horse or mule to satellite forts. Charles Wilkes, leader of the far-flung United States Exploring Expedition sent by President John Quincy Adams, arrived at Fort Vancouver a few years after the Whitmans, in 1841. Among his observations regarding operations at Fort Vancouver, Wilkes noted the presence of barrel churns, which would have greatly simplified the production of larger quantities of butter. Indeed, the fort was capable of producing a significant amount of dairy products: George Simpson remarked on a trip through the area in 1841: "At the dairy [on Wapato Island] we found about a hundred milch cows which were said to yield, on average, not more than sixty pounds of butter each in a year," suggesting the island's operations were capable of producing at least six thousand pounds of butter annually.

We know that the Hudson's Bay Company was making cheese at Fort Vancouver at least as early as Narcissa Whitman's visit in 1836. Fort inventory records from 1838 show that cheese production had become important enough to require an inventory of thirteen cheese vats and one cheese press. Several who visited Fort Vancouver and other HBC forts during this time mention that cheese was produced there in some quantity. Wilkes observed "a large dairy [and] several hundred head of cattle, among them seventy milch cows which yield a large supply of butter and cheese" when he visited Fort Nisqually in 1840. While at Fort Vancouver, he also remarked on the presence of the island dairy operation, noting "two dairies situated on [Wapato Island] in the Willamette where they have one hundred and fifty cows, whose milk is employed . . . in making butter and cheese for the Russian[s] in Alaska." In addition to the more formal operations, some form of fresh goat cheese could well have been made in individual households by French Canadian HBC employees, who would have brought a taste for such cheeses with them.

While no records have yet been uncovered that describe the specific type of cheese made by the Hudson's Bay Company at Fort Vancouver or any of its associated posts, we can gather some clues about the first cheese made in the Pacific Northwest by looking at the state of cheesemaking in England during the same period of time. During the early part of the nineteenth century, the craft of making cheese looked quite a bit different than it does today. At that time, making cheese was primarily a means of preserving milk. The practices and techniques used to make cheese varied greatly by custom as well as by region and were typically passed informally from cheesemaker to cheesemaker, almost always women. Procedures were inconsistent and often improvised. Temperature was ascertained by touch; solidified curd might be cut in smaller or larger pieces, or not at all, depending on the cheesemaker's mood or habit, then pressed and salted according to the custom of each individual dairy.

Despite the relatively primitive state of the craft by modern standards, cheese was produced in great quantities all over England and sent to London mongers by the ton. Cheeses were generally known by the area from which they came (e.g., Gloucester or Lancashire) and the products of a given region were typically similar in style and flavor. Some historians have grouped English cheeses of this period into two broad categories: a northern Cheshire-style cheese that was "of a loose texture and rough austere flavor" and a southern style resembling that made in the Gloucester and Cheddar regions, "milder in taste and of a close, waxlike texture." A few regional specialties, including blue-veined cheeses from Stilton, were also beginning to gain widespread attention and appreciation.

While we don't know a great deal about HBC's dairy workers, it is reasonable to assume that the dairy products they produced resembled those

The Capendales' Short Tenure at Fort Vancouver

In 1835, John McLoughlin's superiors in London informed him they had hired two people to assist with the agricultural efforts at Fort Vancouver:

> In order to give the farming establishment a fair trial, we have engaged a well informed practical agriculturalist William Capendal[e] and his Wife; the one as Bailiff and the other to manage the Dairy Department, they bear most excellent characters and have been strongly recommended to us as thoroughly understanding the breeding and management of cattle of every description [.]

The Capendales were sent from the estate of Sir John Pelly, Governor of the Hudson's Bay Company. Pelly's estate in the southeastern part of England included extensive farming operations.

Mrs. Capendale's position as dairy manager was not unusual; in fact, during this period in England's history, the dairy was considered entirely a woman's realm. According to a contemporary treatise on farming, "the duties of the dairy-maid are well defined . . . her principal duty is to milk the cows and manage the milk in all its stages, bring up the calves and make into butter and cheese the milk that is obtained[.]" HBC's dairy at Cowlitz Farm was also run for some period by an English woman, whose husband labored as a "kind of foreman about the farm."

As it turned out, the Capendales' stint at Fort Vancouver was short-lived. No doubt Fort Vancouver's grounds, carved out of the untamed wilderness, were in stark contrast to the more organized state of farming operations back in England. George Roberts, a clerk at Fort Vancouver at the time, put it more succinctly: "[the Capendales] had been accustomed to high farming and were quite out of their element in a new country." Adding insult to injury, a house that was to be built for them at Fort Vancouver never materialized. John McLoughlin noted somewhat defensively in a letter back to headquarters that "it is impossible [for us] to build them a house without neglecting more important work," emphasizing the fact that he himself did not yet have a house.

By November 1836, McLoughlin wrote to his superiors: "I dare say you will be surprised to see Mr. Capendale back to England," noting that "[his] wife finds things she says, different than what she expected."

they were familiar with from their home country. Several were sent to the Pacific Northwest from the estate of Sir John Pelly, HBC governor (essentially its CEO) during the company's years of operation in the region. Pelly's estate was located in Upton in the county of Essex in southeastern England. Putting the pieces of the puzzle together, we can surmise that the Hudson's Bay Company's cheeses probably resembled Gloucester or Cheddar cheese.

The Decline of English-made Cheese in Oregon

Despite the attractive business model the Pacific Northwest represented during this period—with seemingly endless natural resources there for the taking—for many decades the hardships associated with exploiting those resources outweighed the potential rewards. The rugged, remote landscape on the far side of the Rocky Mountains was difficult to reach both by ocean and by land until the advent of transcontinental train travel much later in the nineteenth century. The nomadic nature of the fur-trapping lifestyle, turf wars between competing commercial and national interests in the region, as well as ongoing conflicts with the indigenous peoples discouraged settlement and commercial development. In retrospect, it's no surprise that it took a well-established outfit with deep pockets like the Hudson's Bay Company to reap significant economic benefit from the area.

Meanwhile, east of the Mississippi River in the burgeoning United States of America, a growing migratory and expansionist sentiment was working its way into the collective national psyche. The developing *zeitgeist* grew out of a number of factors, not the least of which was the Louisiana Purchase of 1803, which extended U.S.-owned territory by over 500 million acres into current-day Montana, Wyoming, and Colorado. Tantalizing accounts of what Americans referred to as "Oregon Country" brought back by Meriwether Lewis and William Clark, who had reached the Pacific Ocean on their trek across the continent in 1804-06, further fanned the flames. The United States' growing collective sense of manifest destiny had the effect of pushing hearts, minds, and covered wagons westward despite the multiple difficulties such a journey entailed. The eventual influx of thousands of white settlers traveling over the Oregon Trail would introduce a new era of agricultural production to the Pacific Northwest.

Some of the earliest non-HBC-affiliated whites to travel to Oregon Country were missionaries riding a wave of evangelical religious fervor that swept across the United States during the early nineteenth century. Among the first to travel west was Methodist minister Jason Lee, who journeyed to the region in 1834 with an expedition led by would-be fur trader Nathaniel Wyeth. Presbyterians Marcus and Narcissa Whitman followed soon after, in

1836, traveling with a large party of fur traders on their way to the annual fur-trading rendezvous in what is now Wyoming. The Whitmans established a missionary outpost at Waiilatpu near present-day Walla Walla, Washington. Among those traveling with the Whitman party were Henry and Eliza Spaulding, who established the Lapwai mission to the west of the Whitman's establishment near what later became Lewiston, Idaho. The earliest Catholic missionaries came to the area initially at the behest of the Hudson's Bay Company to minister to its many Catholic French Canadian employees. Fathers Blanchet and Demers established the Pacific Northwest's first Catholic church and mission at Cowlitz Prairie in 1838 and another a few years later at St. Paul, Oregon, in the Willamette Valley, then home to many ex-HBC employees. During the 1830s and 1840s as many as thirty missions of diverse denominations busily proselytized all over Oregon Country.

Regardless of their religious persuasion, agriculture was key to the missionaries' survival. Most obtained supplies in order to get started, including seed and cattle, from Hudson's Bay Company outposts including Fort Vancouver, Fort Colville, and Fort Nez Perce. So equipped, Methodist missionaries led by Jason Lee's brother, Daniel, at Wascopam near what is now The Dalles, in Oregon, managed to plow twenty acres and plant peas, wheat, potatoes, and corn during their first spring along the Columbia River. That crop didn't fare well, however, as the nascent farmers "barely recovered our seed" from the wheat and pea crop, and the Indians stole the corn and most of the potatoes. Clearly frontier farming was easier said than done; many missionaries were ill equipped for long-term success in farming, lacking not only the tools but also the skills necessary to cultivate the land successfully. Marcus Whitman had better luck at Waiilatpu. Whitman had purchased seed prior to his journey west and tilled sixteen acres around the mission by the spring of 1837. His efforts that year were rewarded with, according to wife, Narcissa, "a substantial harvest of about 250 bushels of potatoes, two hundred bushels of corn and an abundance of garden vegetables." The Whitman compound eventually grew to boast additional wilderness luxuries such as a grist mill and a resident blacksmith.

As was the case for the fur traders, cattle were an important survival asset for the missionaries. The Whitman party managed to bring seventeen cattle, including four milk cows, with them on their westward journey and enjoyed fresh milk along the entire route. Narcissa Whitman used the milk, among other things, to make bread for her traveling party. Once the Whitmans had arrived and were setting up their household on the banks of the Columbia River (a bit too close, as they later discovered during a flood), they were faced with the same dilemma John McLoughlin had faced years earlier—to have any hope of propagating a herd, they could not butcher any cattle. As a result, beef was not served on the Waiilatpu mission tables for five years and missionaries subsisted on horse meat instead.

Cows provided milk and it fell to enterprising mission wives to preserve the precious asset by turning it into butter and cheese. Jason Lee's wife, Anna Maria Pittman, who bravely traveled by ship from Boston to the Willamette Valley in 1837, immediately set to work in the mission kitchen. "I have made twelve pounds a week of butter since I have been here," she remarked in a letter written a few months after arriving. George Simpson of the Hudson's Bay Company passed through the Wascopam mission at The Dalles during a tour through the Columbia District in 1841, mentioning in passing a dairy at the mission with "an adequate supply of milk and butter." Narcissa Whitman, who had been very taken with her tour of the dairy at Fort Vancouver, made butter and cheese regularly once established at Waiilatpu. Marcus Whitman boasted about his wife's skills to his brother in a letter: "Tell mother we are eating cheese of Mrs. Whitman's make, that milk and butter are abundant with us and so will cheese be if we choose to make [more of] it." Mary Walker, who with husband, Elkanah, operated the mission at Tshimakin, north of present-day Spokane, Washington, among the Spokane Indians, regularly listed among her daily chores the tasks of churning milk for butter and making cheese.

Lacking most of the comforts of home, mission wives improvised in many aspects of daily life in the wilderness. To make cheese, the missionaries needed rennet. But since cattle's lives were mostly spared in the early years of mission life, no rennet was available for use in making cheese. John McLoughlin solved this problem at Fort Vancouver by sacrificing just one or two bull calves for rennet. But lacking the far-flung supply network of the Hudson's Bay Company, missionaries were understandably reluctant to part with even one calf. As Marcus Whitman put it, "Calves rennet is a scarce article, for we value a calf as the same as an old cow or ox for it costs nothing to raise them." Yet the missions did not do without cheese; several of the wives used rennet derived from deer as a substitute, an idea conceived of by Mary Walker. She convinced an Indian to bring her the stomach of a young deer, which turned out to work quite well as a substitute rennet source for mission cheesemaking purposes.

The Catholic Church had established a strong presence on the west side of the Cascade Mountains, ministering to Hudson's Bay Company employees in the Willamette Valley and Puget Sound areas; beginning in the 1840s, Jesuits under the stewardship of Father Pierre-Jean DeSmet developed a network of missions on the other side of the mountains through areas we now know as eastern Washington, northern Idaho, and western Montana. Much like George Simpson of the Hudson's Bay Company had decades earlier, DeSmet journeyed westward with fur traders in 1840 to assess the prospects for establishing missions in the region and found the Native populations eager to receive the blackrobe leader (so-called because of the Jesuits' flowing black clerical garments). After returning to his Missouri headquarters, DeSmet embarked on the first of several goodwill-building and fundraising campaigns in the United States and Europe, the fruits of which underwrote the eventual establishment of over a dozen Jesuit missions in the Pacific Northwest in the latter half of the nineteenth century.

Like their Protestant counterparts, Jesuits developed agricultural operations at a number of their regional missions. Isaac Stevens, railroad surveyor and first territorial governor of the state of Washington, visited the Sacred Heart Mission near present-day Coeur d'Alene, Idaho, in 1853 and painted a brief portrait of the mission farm. "[The mission] stock consists of about twenty cows, eight pairs of oxen and ninety pigs which are driven to pasture upon the prairie by Indian boys daily." Brother Charles, who for a period oversaw mission operations amongst the Coeur d'Alene Indians, "attends to the indoor work, cooks, makes butter and cheese, issues provisions and pays the Indians for their work." The St. Ignatius mission near Fort Colville also maintained a sizeable farm, with one hundred sixty acres under active cultivation, a significant herd of livestock, and a dairy that produced both butter and cheese.

Inevitably, tensions developed between the Hudson's Bay Company and the ever-increasing population of missionary and other non-company-affiliated settlers. HBC lacked both the will and the means to actively police those who were beginning to make their way to the region. At the same time, the company anticipated future difficulty controlling the territory George Simpson had hoped to claim for the British Empire if significant numbers of non-HBC settlers and/or Americans began populating the area (a fear that was in fact eventually realized). Loath to tolerate threats to its territorial and economic dominance and still intent on claiming territory north of the Columbia River, HBC made some attempts to discourage interlopers. One means by which it accomplished this was through cutthroat competition—simply paying a higher premium for animal pelts than any potential competitors could hope to. The company also brought in retired operators from its North American headquarters at Hudson's Bay to settle in the region out of an apparent belief that, if it could successfully populate the territory with its friends, it could keep out its enemies. But perhaps the most important way that HBC actively worked to discourage outside settlement was by keeping a very tight hold on that most significant of sources of survival in the remote wilderness—cattle. While John McLoughlin was known for his generosity towards travelers in the region, his practice was to lend rather than sell cattle and supplies to missionaries and other settlers trickling into the area, leaving them perpetually in debt to HBC. This was the genesis of several significant cattle drives that brought large numbers of non-HBC-owned cattle to the Willamette Valley in the early nineteenth century.

The Willamette Valley Cattle Company was formed by a group of settlers that included missionary Jason Lee and Ewing Young, a former fur trader who left the business and settled in the Willamette Valley. Young and his men traveled in 1837 to the Spanish-held territories of the desert Southwest, where they purchased over six hundred cattle and drove them from Mexico through California and north into the Willamette Valley. In 1840, a second group led by Joseph Gale mounted a dramatic effort to bring an even larger group of cattle from California via a ship built solely for the purpose that they called the *Star of Oregon*. The group sailed to San Francisco, sold the ship, and spent the proceeds on a substantial number of cattle, sheep, horses, and mules that they drove overland back to Oregon in 1843.

Just as Spain's introduction of cattle to the Pacific Northwest at Nootka Sound in the eighteenth century had initiated a profound paradigm shift in the region, these later waves of imported livestock set into motion another round of changes that would eventually see HBC out of business and its holdings liquidated south of the 49th parallel. Independently owned cattle were instrumental in enabling the growing numbers of settlers in the Pacific Northwest to sever their dependency on the Hudson's Bay Company. In fact, the fur business was already on the decline during the mid-nineteenth century, due in part to sagging worldwide demand for fur and in part self-generated by the near extinction of various fur-bearing species like beavers and sea otters. In 1846, the British and Americans finally resolved the territorial power struggle that had been simmering just below the surface since the Joint Operating Agreement of 1818 and formally divided up the territory of the Pacific Northwest at the 49th parallel, which today forms the international boundary between the United States and Canada. The resolution was a green light to United States territorial expansion and white settlers began to swarm into Oregon Country.

Though HBC's operations continued in the United States for some years after, the 1846 treaty effectively put the company out of business south of the 49th parallel. In 1849 HBC formally relocated its headquarters from Fort Vancouver to Fort Victoria on Vancouver Island (built in 1843 in anticipation of just such an outcome) and continued its trading operations to the north. HBC-owned forts and farms south of the new international border fell into decline and were eventually displaced by individual American homesteaders. Most HBC-owned cattle around Fort Vancouver either dispersed (roaming herds of aggressive wild cattle were a common problem during this period) or were appropriated by the remaining Fort Vancouver employees. By then HBC's valuable land on the north bank of the Columbia River, just opposite the developing commercial center of Portland, Oregon, was the frequent target of squatters. The United States Army took over Fort Vancouver in 1849 and maintained it as a military outpost for over a century until ceding the site to the National Park Service in 2011.

HBC's Puget Sound Agricultural Company fared no better after 1846. Non-company settlers routinely disregarded property lines around both Cowlitz Farm and Fort Nisqually in what later became Washington Territory, vandalizing buildings and moving company fences to suit their needs. According to a former employee, "[t]here was a strong prejudice in the community [around Fort Nisqually] against the Company." Encroaching settlers shot at and killed HBC cattle freely, though the PSAC did manage to transport much of its livestock northward on ships to Vancouver Island. PSAC finally relocated its center of operations to Vancouver Island in the 1850s, establishing Viewfield Farm, Craigflower Farm, Colwood Farm, and Constance Cove Farm in the region around Fort Victoria that it stocked with sheep and cattle shipped from its former holdings. These new farms continued to supply HBC's remaining operations in the Pacific Northwest north of the 49th parallel as well as fulfilling what was left of the company's contractual obligations to the Russian-American Company. The Vancouver Island operations lost money almost immediately due to mismanagement, lack of labor, and poor farming conditions, and HBC was forced to eventually take over PSAC's assets. In the 1860s the United States government compensated HBC for all of its land and properties, formally ending the HBC era in the western United States.

The impressive agricultural operations that the Hudson's Bay Company had developed and maintained were in many ways ahead of their time, or to put it more precisely, ahead of America's timeline. While the 1846 treaty ended the simmering political and territorial conflicts, HBC's retreat effectively sent western-style agriculture in the Pacific Northwest several steps backward. Fort Vancouver and its satellites practiced European-style agriculture envisioned, created, and financed by a large, centuries-old international company that could well afford to develop a vast trading network (no mean feat in the remote wilderness environment), as well as support and grow the operation by bringing in appropriate resources and equipment. The company successfully produced great quantities of grain, meat, and dairy products that supported the European tastes of the fur trappers who populated the region and fostered the company's ongoing prosperity for several decades. HBC's operations had demonstrated, however, that European settlement in the uncharted reaches of the continent was possible, that the agriculture infrastructure necessary to support such settlement could be established and conducted on a large scale, that crops could be grown and cattle grazed, and that life could in fact resemble more or less what potential colonizers were accustomed to in their home states or countries. Ironically, through its example HBC in some ways enabled its own demise. Though the first cheese made in the Pacific Northwest was English-made cheese, that would soon change.

Before motorized transportation became common, farmers often hauled milk to area creameries and cheese factories by horse. Here Ernest Wheeler (left) pauses with his team en route to Tom Walker's cheese factory in Pleasant Hill, Oregon, in 1910. Photo courtesy Lane County Historical Museum.

Chapter 2
Milk and Cheese in Oregon Country

From a contemporary perspective the scenario sounds a little strange: imagine thousands of people deciding rather abruptly to uproot themselves, their families and livelihoods and travel thousands of miles without the benefit of motorized transportation toward a remote region of the continent most had only heard of or perhaps read about. Odd, perhaps, but such was the trending public sentiment in the mid-nineteenth-century United States. Historians estimate that during the 1840s-60s over three hundred thousand people migrated west over what came to be called the Oregon Trail, cementing the profound changes to the culture and landscape of the Pacific Northwest already set into motion by European oceangoing explorers and the fur trappers that followed them.

The Louisiana Purchase of 1803 and the captivating reports that Lewis and Clark brought back opened the eyes of many U.S. citizens to the existence of a vast landscape full of untapped resources on the western part of the continent. By 1825 legislation was introduced in Congress in support of the extension of United States territory into what had become known as Oregon Country. In 1829 enthusiasts organized a group calling itself the Oregon Colonization Society (later the American Society for Encouraging Settlement of Oregon Territory) in Boston and one of its most vocal supporters, Hall Kelley, penned several treatises strongly advocating national expansion. "No portion of the globe presents a more fruitful soil, or a milder climate, or equal facilities for carrying into effect the great purposes of a free and enlightened nation," Kelley wrote in *A Geographical Sketch of that Part of North America Called Oregon*. While Kelley's efforts inspired a few hardy souls to travel to Oregon Country, increasingly more tangible reasons beckoned opportunity seekers westward. Resolution of the territorial conflicts between the United States and Britain in the 1846 Treaty resolved any lingering doubts about sovereign control. The Gold Rush of 1849 drew attention farther south to what would soon become the state of California, but around the same time gold was also discovered in the sands near Gold Beach (hence the name) in far southwest Oregon as well as in other spots along the southern Oregon coast and elsewhere in Oregon Country. The Donation Land Claim Act of 1850, which offered 640 acres to white married couples who could establish a claim in Oregon Country, along with the Homestead Act of 1862, provided significant material incentive— land ownership. There was no shortage of reasons to go west.

By the mid-nineteenth century a substantial population of cattle, sheep, and other livestock were already in the Pacific Northwest, as we saw in Chapter 1. But that was just the beginning: the mass human migration via the Oregon Trail might as well be called the "Great Nineteenth-Century Livestock Drive" since so many animals (mostly cattle but significant numbers of sheep, goats, pigs, and even chickens) made their way to Oregon and California along with the humans. "Cattle, cattle, it really seems as though the whole country is alive with men, women, horses, mules, cattle & sheep with a smart sprinkling of children," remarked Mary Stuart Bailey during the early days of her cross-continental journey on the Oregon Trail in 1852. The numbers are astounding: in just one year, 1853, the register at Fort Kearney in Nebraska recorded 105,792 cattle, 5,477 horses, 2,190 mules and 48,495 sheep passing through on their way westward. Historians estimate the overall number of cattle and sheep that traveled on the Oregon Trail at over a million.

Cattle were valuable, if not vital, on the Oregon Trail for a number of reasons. They served as transportation (oxen hauled many emigrants' wagons westward, though debates raged over the relative merits of cattle vs. horses vs. mules for this purpose) as well as walking repositories of meat. But despite their obvious value, livestock introduced myriad complications to an already hazardous, arduous journey. Livestock that were not otherwise engaged in pulling wagons were typically herded loosely along the route in groups; travelers devoted much time and energy to ensuring that the animals didn't stray but they inevitably did, causing the humans to devote additional time and energy to recovering them. The animals also drew the attention of ubiquitous wolves and coyotes and were lost to these predators in large numbers along the way.

The physical rigors endured by animals on the more than two-thousand-mile journey were significant. Trail guides devote pages and pages to advice about the proper care of animals' hooves; one advises against taking young calves and foals. "[No horses, mules or oxen] should be taken, which are under five, or over ten years of age; nor should calves or colts, under one year of age, be taken; for, from the tenderness of their hoofs, and their inability otherwise to endure fatigue, they are invariably left by the way." Perhaps most remarkable is the manner in which cattle and other animals would be herded across rivers *en masse*. Losses were common; inevitably, few animals that started the journey actually made it all the way to the West Coast. By the time Cecilia Adams and her party reached the Grande Ronde River in Eastern Oregon, months after starting their journey in Missouri, she lamented: "many cattle are failing and all are very poor and a good many get lost in the thick timber. A good many wagons are left, some broken and some good and sound, because the cattle are not able to take them along." Animal carcasses in varying states of decay were strewn all along the trail route, grim emblems of the hardships of the journey.

Despite the daily struggles of life along the Oregon Trail, many travelers managed not only to bring cows but also to milk them along the way. The *Emigrants' Guide*, one of many guidebooks of the period written for prospective travelers, warns against milking cows on the Trail as the milk would most certainly be unwholesome due to the cow's "exposure to excessive heat, and extreme physical exertion." Joel Palmer, however, advised in his *Journal* that "[e]ach family should have a few cows, as the milk can be used the entire route, and they are often convenient to pull the wagon to relieve oxen." Still, milk was rare enough on the trail to be special. Catherine Haun, whose four sisters had died from tuberculosis and who suffered from the disease herself, traveled west with her husband hoping to be cured by the fresh air and exercise to be found on the journey. Haun wrote of four bachelors in her wagon party who were traveling with four oxen and two milk cows. The men shared the milk from their cows with everyone in their party: "[m]any a cup of milk was given to the children of the train and the mothers tried in every possible way to express their gratitude." Unfortunately Haun's tale grew darker; the milk cows later perished after being forced to pull the wagon after the oxen had perished—but such was the daily life and death drama that was the Oregon Trail experience.

Those who managed to bring along a milk cow and keep it alive incorporated milking into the routine of daily chores. While men typically concerned themselves with feeding and watering animals, maintaining wagons, and hunting for wild game (buffalo, elk, and antelope were plentiful on the plains), milking was most often left to the women and/or children. Delilah Hendershott recalled the routine: "At daybreak, everybody was up to milk the cows, get breakfast, etc. Surplus milk was strained into cans on the backs of the wagons." The task of milking on the Plains was not always easy; one girl had her leg broken by a kick from a cow, but Jean Rio Baker, a Mormon convert who had traveled to the United States from England and was on the overland portion of her journey with the eventual destination of Salt Lake City, reported matter-of-factly in her diary that "[the girl's] father set the bone and she seems to be doing well." And contrary to the dire warnings of the *Emigrants' Guide*, both the quality and quantity of the milk were more directly

affected by the availability of fodder; Helen Carpenter noted in her diary "[t]
here has been such poor feed that Sookey's milk is failing."

Given the severity of the Oregon Trail journey, it would not seem possible for travelers to have engaged in such labor and equipment-intensive
tasks as making butter or simple fresh cheeses, but some did just that. In fact,
Joel Palmer advised that well-prepared travelers should carry "two churns,
one for carrying sweet and one for sour milk." But butter churns were not an
absolute necessity; Helen Carpenter recounted a trick that many other travelers with milk discovered—the rocking motion of the wagon along the bumpy
trail acted as a churn and produced butter as if by magic:

> The milk is carried in a can swung to the wagon bows overhead. By
> noon (if the churn works well and it seldom fails) there is a ball of
> butter the size of a hickory nut and innumerable little ones like shot.
> If the day is hot, we have hot milk; if cold, we have cold milk, but
> unlike the 'bean porrage' of school days, it is never 'nine days old.'

On a rest day, Cecilia Adams kept herself busy baking and making a simple
fresh cheese. "[Did] some washing and I baked bread and pumpkin and apple
pies, cooked beans and meat, stewed apples . . . besides making dutch cheese."

Dutch Cheese

Dutch cheese was a term for a fresh style of cheese made from
soured milk. Commonly made in nineteenth-century households, it
was also referred to as pot cheese or farmer's cheese and resembled
contemporary cottage cheese.

To Make Good Dutch Cheese

Set the sour milk on the stove until the whey separates
from the curd. Then put it into a large, coarse towel or
thin cloth and tie it up. Lay it in a pan and let it lie, occa-
sionally turning off the whey until no more whey runs out.
Then put it in a dish, and with the hands work it in salt to
taste and a piece of butter and cream to make it adhere
sufficiently to make into balls. Some persons like to eat it
with sweetened cream.

—*The New Northwest*, September 18, 1871

Those not lucky enough to have their own cow on the trail were sometimes able to purchase or barter for milk, butter, or cheese along the way. As the numbers of travelers on the trail increased, entrepreneurs began to see economic opportunity in the migrating masses, and trading posts appeared along the route. There, travelers could trade tired animals for fresh ones and/or purchase much-needed supplies including butter and cheese. Several parties that passed through Fort Hall (near present-day Pocatello, Idaho) in 1849 recalled that Mormon families had relocated to the area for the specific purpose of farming and selling their goods to travelers. According to one account, the Mormons had established a considerable encampment in the vicinity of Fort Hall including a herd of three hundred cattle and were making and selling both butter and cheese. "Butter was very scarce as [the Mormons] sold it as fast as they could make it," wrote traveler James Pritchard. "Cheese brought 25 cents per pound." Farther down the trail at The Dalles, Oregon (the end of the land-based portion of the trail until alternate routes were developed around Mt. Hood), Mary Yeargin Laughlin developed a significant business supplying butter, milk, vegetables, and eggs to emigrants who were usually desperately hungry at that late point of the journey. In addition, Laughlin and her husband purchased weakened cattle that managed to survive, fattened them over the winter, and then butchered the cattle and sold the meat to the next crop of arriving emigrants the next year. When Fort Dalles was established on the site of the former Wascopam mission in the 1850s to police increasingly frequent conflicts between emigrants and Indians, Mary Laughlin also supplied butter, milk, and handmade gloves to the resident soldiers.

Farmstead Dairies

Once they arrived in the Pacific Northwest, would-be settlers found an entirely different state of affairs than they'd been accustomed to back home. Millions of acres of old-growth fir, spruce, and cedar covered the more temperate landscape west of the Cascade Mountains. Those dense forests impeded travel and also harbored all manner of predators—including wolves, bears, and cougars—that represented a threat to both settlers and their livestock. The region's numerous rivers and streams were often difficult, if not impossible, to navigate due to unpredictable currents and dangerous rapids and were prone to frequent and dangerous flooding. Of course there were no roads save animal or Indian trails and a distinct lack of bridges, making even several miles of travel an arduous, extended adventure. East of the Cascade Mountains was a vast expanse of even less hospitable arid desert.

Travelers who survived the arduous journey to Oregon Country soon found that the determination, perseverance, and self-sufficiency skills drawn upon while traveling westward would be exercised repeatedly once they arrived. Like the fur trappers and missionaries before them, the settlers' first task upon starting over in the West was establishing a home. Those with the stamina and wherewithal registered a homestead claim in the hopes of being able to establish ownership by maintaining and cultivating the spot for at least five years—no mean feat. But even with the possibility of land ownership on the distant horizon, the immediate practical reality of the situation was challenging at best. Even for families with some means, all the money in the world could not buy goods that simply did not exist in the remote region. Many simply continued to sleep in their wagons until they could gather or purchase materials to construct a shelter. Joseph Champion, the first white settler in Tillamook, Oregon, in 1851, famously lived in an enormous hollow tree stump until he could build a house—actually not an uncommon practice in early settlement days, in this land where old-growth trees could exceed twenty feet in diameter. Hopefully the homestead site of one's choosing was near a source of fresh water, otherwise the days would be filled with the backbreaking task of digging a well or carrying water great distances. Homes were typically built out of logs cut from trees settlers would of necessity fell themselves; hewn lumber was a luxury possible only once sawmills had been established. Those aspiring to cultivate the land faced the overwhelming task of clearing the ubiquitous trees, dense brush, and rocks first.

After shelter came the question of sustenance. Settlers with cattle or the means to purchase them found themselves at an advantage since they were in a position to feed themselves as well as make money. Silas and Lydia Plimpton built a herd of sixteen cows at their homestead claim in Rainier, Oregon. Lydia used the cream to made butter which she sold for fifty cents a pound; she reported in a letter to her mother in 1856 that "we find a ready market for all the butter I can make . . . [and I] have sold milk for the last three months at 50 cents a gallon." Plimpton also told her mother that her husband "talks very strong" about getting into the cheese business, but the prospects for that seem to have depended on how long they remained in the area. Martha Gilliam relates a similar story; her family traveled over the Oregon Trail and settled in the Willamette Valley near Dallas. Her father later led a regiment sent to battle the Indians in the Walla Walla area after the Whitman Massacre. After their father's death in battle, Martha and brother Marcus took over farming chores; she recalled milking twenty-four cows during the summer of her fourteenth year, 1853—Spanish cattle because the family could not afford the more expensive European breeds—and she made and sold butter at fifty cents a pound. She and her brother saved $800 that summer selling butter and bacon.

Lucinda Collins Fares. Photo courtesy Snoqualmie Valley Historical Museum.

Lucinda Collins' family was among the first group of settlers in what would eventually become King County, Washington; Lucinda and her mother are said to have been the first white women to arrive in the area. Lucinda married ex-Hudson's Bay Company employee Joseph Fares and moved to the Snoqualmie Valley near present-day North Bend, Washington, in the 1850s, the first white woman to settle in that region as well. The Fares kept and milked a herd of about thirty dairy cows on their farm, and Lucinda was a natural with the bovine breed. Because she found it difficult to walk out to the pasture and milk each cow herself, she is said to have called her cows by name at milking time and they came to her: "'Pidey,' she would call out, and then Pidey would come over obediently to be milked. When Pidey had given up her milk she would pass on and then Lucinda would call Mary, and so on, until the thirty were milked." Fares became well known as a local purveyor of butter.

Several early pioneers in the region now called the Olympic Peninsula in Washington also found success peddling their dairy products to fellow set-tlers and the surrounding community. Englishman William Bishop arrived in the Pacific Northwest as a sailor on the British frigate *Monarch*. Bishop and several others jumped ship while the *Monarch* was docked in Victoria, just to the north and within sight of the United States coastline, and made their way to present-day Chimacum, Washington. Bishop acquired land and began a dairy farm that proved quite successful; he sold milk, butter, and eggs to local residents and to settlers of the nearby towns of Port Ludlow and Port Discovery in the 1850s and '60s. In the early 1870s, Alonzo Davis brought a small herd of Jersey cows with him from Canada to what is now Clallam County, Washington, near the town of Dungeness. Davis developed a thriving business selling his Gold Edge brand of butter to settlers and lumber camps in the area and to markets around Puget Sound.

Farmers who settled near more populated areas benefited from an established population hungry for the dairy products they were accustomed to enjoying back home. The *Idaho Tri-Weekly Statesman* was impressed enough by farmer Ed Bryon in 1874 to call him "a capital butter and cheese maker," and advised readers where they could buy his products in the Boise area. Another early entrepreneur, Mrs. Fred Millman, sold small three- to five-pound wheels of cheese from her Boise home in 1876, garnering a mention from the local newspaper because "they are just the right size for family use and will not dry and spoil as is the case when the same number of pounds are cut from a large cheese."

Portland, the largest city in the Pacific Northwest until the early twentieth century, attracted a number of farmers and dairymen. Andrew Jackson Dufur, Sr., left Vermont with his family and arrived in Oregon during the 1850s. He first leased and then purchased the former Quimby Farm north of Portland and established a dairy there. His "Dufur's Premium Cheese" was advertised heavily in the *Morning Oregonian* newspaper throughout the 1860s. One enthusiast reported that Dufur's cheese was "equal to English cheese" (meaning very good) in quality. Dufur went on to become a state legislator and traveled as a state delegate to the Great Philadelphia Centennial in 1876. Two of Dufur's sons later went into the ranching business east of Portland in Wasco County, and A. J. Dufur, Jr., continued the family tradition, raising dairy cattle and making cheese at his ranch there.

Astoria, at the mouth of the Columbia River, grew into a thriving port city and workers at its many lumber mills and salmon canneries provided a ready market for dairy products. The Clatsop Plains, flat prairies on the south side of the Columbia River near present-day Warrenton, were particularly conducive to cattle grazing and thus also to dairy farming. As early as the mid-1840s a number of the earliest settlers had established small farms there and Alva Conditt had started a dairy and was making cheese. Sarah Owens, who traveled to Oregon Country in 1843 with a party that included Jesse Applegate and family, was said to have been producing dairy products during the 1840s, likely butter. During the 1860s, Owens' daughter Bethenia collaborated with area farmer John Hobson to produce a large cheese to send to Union troops fighting in the Civil War. Hobson constructed a cheese hoop, a large mold in which the cheese curds were pressed and formed, out of part of a whiskey barrel and area farmers contributed the necessary milk. The cheese, which Owens described as "mammoth" in proportions, was sold for $145 at auction in Astoria and then later conveyed to the Oregon State Fair in Salem, where organizers auctioned it off again to benefit the war effort. According to Owens, "whether the cheese itself ever reached the 'boys in blue' and whether or not they found it palatable, and digestible, I was not informed." Bethenia Owens-Adair went on to gain prominence as one of Oregon's first female physicians.

Those who settled in more remote, unpopulated regions of the Pacific Northwest faced significant challenges in finding a market for their wares. The central Oregon coast presented myriad hardships for new settlers; it's a wonder that any managed to make the trip over the Coast Range, the mountain range that separates the Oregon Coast from the Willamette Valley, prior to the existence of roads. When Warren Vaughn set out to visit the Tillamook area in 1852 with four other men, all with an eye toward settling there, the party lost their way and eventually arrived at the Pacific Ocean forty miles south of their destination. A second try proved more successful. Though difficult to access, the area had one significant asset—plenty of green grass and a mild climate that allowed it to flourish most of the year, ideal conditions for grazing cattle. According to Vaughn, cattle had already been herded to the Tillamook Area from the Clatsop Plains by David Wilson in 1852. By the time of Vaughn's arrival several families were already living in the area and both the Doughertys and Trasks already had dairy cows. Vaughn eventually acquired Wilson's cattle and began milking them regularly. Because the region was so remote, cattle were central to these settlers' survival; Vaughn and others regularly made dutch cheese, which supplemented their otherwise steady diet of salmon and occasional garden produce. Eventually the hardy settlers banded together, felled the timber that was all around them, scavenged iron fittings from the many shipwrecks on nearby beaches, sealed the seams between wood beams with rope, and salvaged canvas to make sails and managed to produce a ship, the *Morning Star*, which they used to transport their butter and other goods to markets in Astoria and Portland.

The problem of access to markets was often a considerable challenge in the early days of settlement in the wilderness. Jock Morgan and Civil War veteran William Kiester were among the early white settlers in the Kittitas Valley on the east side of the Cascade Mountains in what would eventually become the state of Washington. Both milked cows and made butter, which they transported south to, and then across, the Columbia River to sell in The Dalles—a seven-day journey. In order to ensure that their precious cargo survived the trek, the farmers packed the butter in barrels or firkins lined with cheesecloth, spread a one-inch layer of salt over the top, and then wrapped cheesecloth over the top of the salt layer; the barrels were then covered with blankets for the journey. In Lane County, Oregon, near Eugene, Sydney Stephens kept a herd of about ninety cattle and made cheese that he periodically transported by the wagonload seventy miles to Florence on the coast to sell. But the enterprising Stephens made the trek worthwhile; at Florence he would pick up a load of salmon that he sold along the way back, making the sales trip profitable in both directions. According to son Clair, Sydney Stephens also hand-peddled his cheeses all over the surrounding region. In later years, as the population and markets grew, he began to sell it through local merchants. John Hansen of Cottonwood Creek, east of Twin Falls in what is now southern Idaho, faced similar challenges in transporting his goods to market; he managed to sell most of his cheese through the general store in Kelton, Utah, over one hundred miles away. Hansen's operation didn't last long, however, as he and his wife were forced to flee the area due to ongoing conflicts between the area's white settlers and the native Bannock Indians.

The many mining and lumber camps that multiplied across the Pacific Northwest became captive markets that settlers learned to exploit. In Marion County, Oregon, Cynthia Ann Applegate, wife of Jesse Applegate of Applegate Trail fame, took advantage of the cattle the family brought over on the trail to the Willamette Valley in 1849. According to daughter Sallie Applegate Long she was "an expert butter and cheese maker" and developed a lucrative business sending her cheeses to the region's mining camps. Applegate milked all of her cows herself "for no one could do it just right—and she loved her cows as though they were human beings." When the family later moved to Yoncalla in the Umpqua Valley in southern Oregon, Cynthia Applegate continued to ply her trade as Yoncalla was conveniently situated near a supply route than ran through the region. By Long's account, Jesse Applegate would often return from long surveying trips with little surprises to lighten his wife's labor, including dairy equipment such as churns, cheese hoops, and presses. In a similar fashion, James and Louisa Bennett used the milk from their herd of seventy-five cows to make butter and cheese that they transported by the wagonload to sell in the mining communities of Idaho City and Placerville during the mid 1870s. Benjamin Boswell, considered to be one of the first dairymen in what would later become Wallowa County in eastern Oregon made more butter and cheese during the 1880s than he was able to sell to neighbors. He sought additional markets in the town of La Grande (today a sixty-mile trip on a comfortably maintained state highway) as well as at the nearby Cornucopia Gold Mine. Many entrepreneurial settlers quietly earned an income supplying all manner of farm produce, milk and meat to the ubiquitous mines and lumber camps across the Pacific Northwest.

Early butter and cheese production in the Pacific Northwest evolved out of the stark necessity of survival. Those with cattle that survived the Oregon Trail journey, or those with the means to purchase cattle once they arrived, fed themselves in the short term by butchering the cattle, or longer term by milking them. Having a dairy was as simple as possessing a cow and a pail and perhaps a wooden butter churn, though the latter would not have been an absolute necessity. Because fresh milk could not be preserved for any significant length of time before the advent of refrigeration, it would have been crafted into butter or cheese as a matter of course. A family's sustenance might come either directly from the nourishment of the milk, from the source of income it provided, or both. Across the Pacific Northwest, early white settlers developed localized micro-economies in which individually produced or grown farm products were a medium of exchange, a pattern of informal trade they'd grown accustomed to while traveling westward on the Oregon Trail. These earliest homesteaders and homemakers—most of whose names we will never know—were the first regional cheesemakers of the post-HBC era.

The Rise of an Industry

The production of dairy products increased steadily during the latter half of the nineteenth century in the Pacific Northwest. As early as 1850, Oregon Territory (which included all of what would later become the states of Oregon, Washington, and Idaho) produced over 211,000 pounds of butter and 36,000 pounds of cheese, and the figures are almost certainly low because accurate information would have been difficult to come by during this period. Total recorded output increased to nearly two million pounds of butter and 97,000 pounds of cheese region-wide by 1870 and continued to grow quickly in all three Pacific Northwest states over the next several decades despite a series of economic depressions in the 1890s that drove down prices for dairy products significantly. The rapid expansion maps an increasingly organized mode of dairy products production intended for a market beyond family and neighbors.

Historically, the consumption of butter exceeded that of cheese consumption in the United States. In 1909, the earliest year for which figures are available, butter consumption averaged over 17 pounds per person per year while the rate of cheese consumption was considerably lower, at about 3.5 pounds per person per year. "Americans taste cheese, while Europeans eat it," noted Henry Alvord, a prominent dairy scientist of the day. Alvord pointed to high prices, poor care of cheese on both the manufacturing and consumer side (dried or otherwise spoiled cheese being unpalatable), as well as the generally bad quality of available cheese on the domestic market as among the reasons for Americans' lack of enthusiasm about cheese. For early Pacific Northwest farmers, butter making would have been advantageous for the average person because it required less skill and equipment than did making cheese, for which some understanding of specialized techniques and processes was required. In addition, butter could be made and sold for an immediate profit as it didn't require aging time to perfect as did cheese. At the same time, cheese brought higher prices and could be stored longer than butter, so cheese production did possess distinct advantages that some farmers were able to exploit.

As settlers continued to stream into the region and the local economy grew, farmers with the skill to produce consistently good quality cheese became known for their wares. Among the cheesemakers who rose to local prominence in the Willamette Valley was Warren Cranston. Originally from Illinois, Cranston traveled over the Oregon Trail with his father and first wife and a number of others in 1851. According to C. K. Cranston, his father started to make cheese on his farm in the Waldo Hills area (near what is now Silverton, Oregon) in the 1860s. Like many of the earliest settlers, Cranston first made cheese primarily for his family, but he quickly developed a reputation and began to sell to neighbors who stopped by the farm requesting his cheese. Retailers in Salem, the state's capitol, saw an opportunity and began selling Cranston's cheese as local demand grew. "Cranston's Cheese has become a family byword among Oregonians," a local newspaper proudly proclaimed. He made a reported thirty thousand pounds of cheese in 1894. Unfortunately no records have surfaced detailing the specific type of cheese Cranston made or how it may have tasted, though his cheeses won multiple awards at state competitions. Cranston went on to become a member of the Oregon State Legislature; one pundit opined that Cranston's "Republicanism is as creamy as his cheese," an apparent reference to his trustworthy nature. According to his son, others took over the cheese operation after his father's death in 1899, but the once celebrated Cranston's Cheese went the way of its founder.

Another Willamette Valley farmer had the means to develop dairy and production facilities on a significant scale. Henry Ankeny, son of prominent Portland banker and developer Alexander P. Ankeny (after whom many prominent Portland landmarks are named), operated one of the largest dairy farms in the Pacific Northwest on his 4,500-acre estate south of Salem,

Oregon, during the 1870s and 1880s. Using the milk from his herd of over one hundred cows (an enormous herd for the period), Ankeny's hired help made cheese during the summer months and butter during the winter season. The cheese room was outfitted with state-of-the-art equipment including Roe's Patent Premium Cheese Vat and Heater and ten cheese presses. The curing room must have been a sight to behold with hundreds of wheels of cheese resting on shelves two deep. Local publication *Willamette Farmer* noted: "those who tasted the premium cheese at the state fair last year need not be told that Henry Ankeny makes good cheese."

By the 1880s, a thousand cows inhabited an area along the Oregon coast stretching from Astoria all the way to Nehalem Bay, about forty-five miles to the south. The region's most prominent dairy farmer was Josiah West, who kept a milking herd of seventy-five Holstein, Jersey, and Ayrshire cows. West's Clatsop Factory Cream Cheese (the term "cream cheese" used to distinguish it from cheese made with skim milk) was produced in two sizes, in wheels of thirty pounds and of eight pounds more convenient for household use. One appreciator described West's cheese as "the most delicious I have ever tasted." One of Josiah West's sons went on to make cheese for another area dairy farmer, Mr. Carnahan, for several years. Dairy farmer Charles Ward had learned to make cheese in California before relocating to Clatsop County, first making cheese at a farm owned by the Wingate family before purchasing his own on the Clatsop Plains. A Mr. Butterfield produced Butterfield's Clatsop County Factory Cheese, albeit reportedly in relatively small quantities. A contemporary publication attributed the area's cheesemaking success in part to its proximity to the ocean, "which ensures during the summer season a naturally suitable temperature in the storerooms for excellent curing of the cheese."

Dairy farmers just across the Columbia River from Portland in Washington Territory took advantage of that growing urban marketplace. The Hathaway family made cheese in the Union Ridge area (now Ridgefield) that was marketed in Portland in the early 1870s. Another of Clark County's early cheesemakers, C. T. Stiles of Washougal, a territorial legislator, reportedly produced as much as eighteen thousand pounds of cheese in 1873. One newspaper waxed enthusiastic about Stiles' operation, remarking: "he has a superior facility for [the cheese] business and is turning out a fine article." A visitor to Stiles' farm in 1877 noted that Stiles was milking sixty-three cows. Nearby, Mr. Cramm of Brush Prairie was milking about twenty-five cows and making one hundred and seventy five pounds of cheese every week.

Henry Koch started a creamery in Woodland in the spring of 1887, hiring Henry Hanna as his butter and cheese maker. When the price he could get for cheese exceeded that for butter, Koch converted his operation to a full-time

The Cheese Heist of 1889

Evening Capitol Journal, Salem, Oregon, Monday October 7, 1889

A CHEESE CASE—Quite a sensational arrest was made out beyond the depot this morning by Deputy Sheriff Savage. Some time ago Fred Luthy started a cheese factory at Stayton, and as he was rather pushed in financial matters, the good farmers in that section agreed to sell him milk with the understanding that he was to pay for same when the cheese were sold. He made quite a number of cheese and was selling it. The farmers thought he was not doing the right thing with them. About a month ago he moved to Salem and started a boarding house. Two weeks ago J. W. Thomas and C. W. Thomas decided that all was not right and sued Luthy for $450.80 due them for milk and obtained Judgment. They then attached his goods in the cheese factory, about three miles east of Stayton, and stowed them away in a cellar belonging to J. W. Thomas, about three miles east of Stayton. Nothing went wrong until this morning, when J. W. Thomas went to salt the cheese, which he is required to do every other day and to his great surprise found that the cellar had been broken into and a load of cheese stolen. He at once went to Stayton, notified Constable J. P. Thomas, who set out for Turner on trace of the stolen property. When he reached Turner he learned that the parties had passed through there and for fear he would not catch them he telegraphed Sheriff Croisan, who at once sent Mr. Savage to meet the gents. He captured them just beyond the depot and promptly arrest them. They proved to be F. Buech a brother in law of Mr. Luthy. Buech claims to have a bill of sale from Luthy and Alplanalp was only hired to haul the cheese. They were arraigned before Justice Chase and lodged in the county jail until tomorrow for trial. They went to the cellar at 3 o clock this morning, it seems, and loaded up forty or fifty cheese, and started to the city with the above results.

cheese plant. Unfortunately several years later the price of cheese plummeted. This forced Koch to significantly lower the price he paid to farmers for their milk, since he'd sold all of his butter-making equipment. Koch was already not well liked among the dairymen of the area and his actions deepened their distrust; the farmers eventually refused to sell milk to Koch. After several attempts to buy him out failed, the dairymen became determined to start their own cooperative plant instead.

One of the newly formed Woodland Dairy Association's best business decisions was to hire cheesemaker Peter McIntosh. McIntosh, who had learned the trade in his native Ontario, was at the time operating a cheesemaking plant he had started in nearby Freeport (today part of Longview, Washington). McIntosh helped the Woodland farmers open the Woodland Co-Operative Cheese Factory and became its first cheesemaker when the facility opened in 1889. No doubt the farmers were pleased that Koch was subsequently forced out of business due to lack of milk. McIntosh eventually left Woodland in 1894 to work for T. S. Townsend and Harry Ogden at their new plant in Tillamook, Oregon. Unfortunately after McIntosh left the quality of cheese began to suffer greatly at Woodland—it was said to be soft and of poor flavor—leaving the operation in grave danger of going under. The farmers called upon McIntosh once again for help. McIntosh not only arranged for a Mr. Daniels to sell the poor quality cheese to Seattle merchants (which the enterprising salesman did for a remarkable 2 cent per pound profit), he also arranged for his brother-in-law John Bogart, himself a skilled cheesemaker, to travel to Woodland from Ontario, Canada, to run the factory. Business rebounded and the Woodland plant grew to become one of the larger cheese plants in the state of Washington by the turn of the century.

Because the Pacific Northwest was growing and developing and so many goods, food and otherwise, were of necessity being imported into the region, a class of entrepreneurs began to view the region's nascent dairy industry as a good business proposition. What was likely the earliest organized cheesemaking operation on the Olympic Peninsula was founded in Chimacum, south of Port Townsend, in 1878 by one Mr. Fallows of Tacoma. By April of that year, the local newspaper trumpeted enthusiastically: "[t]he North Pacific Cheese Factory is in full blast, turning out an excellent article of cheese in large quantities, the amount being in the neighborhood of five hundred pounds per day." In May the factory was shipping cheese to Victoria, Seattle and Portland. The factory closed in September for the season with promises to reopen the next year, though buried in the same edition the newspaper reported rather ominously "rumor has it that some changes will be made in the ownership and management of the enterprise."

To the south in Lewis County, J. Henry Long (later a state legislator) and Canadian cheesemaker William Birmingham founded a cheese factory in Claquato near Chehalis in 1877, a smart prospect given that the Northern Pacific Railroad had begun operations in the area in 1873, connecting nearby Chehalis with the Puget Sound area's lucrative markets and ports. A reporter toured the plant in 1878 and was particularly awed by the aging room: "[h]ere we found upwards of six hundred cheeses of various sizes and in different stages of perfection, each neatly stenciled with the date of manufacture presenting in their rich golden color a sight that would arouse the longings of the most dainty epicure." The plant took in milk from seven area dairy farmers as well as from Long's own herd of fifty cows. Long and Birmingham also operated two other cheese factories, one in the Sumas Valley just over the border in British Columbia and a third on the White River near Kent. Henry Hanna, who went on to become the cheesemaker for Henry Koch in Woodland, got his start in the cheese business at Long and Birmingham's Claquato plant.

Entrepreneur Thaddeus (T. S.) Townsend, who came to Oregon from Iowa with creamery management experience, recognized the potential for dairying and cheesemaking in the Tillamook area. He traveled to the region from Portland in 1893 and enlisted the help of area farmers, offering to start a factory if they would agree to provide the milk. The farmers were amenable to the idea, but since there were not enough dairy cattle in the area to justify starting a cheese factory, Townsend brought additional cattle over himself. He drove a herd of twenty-five cows over the Coast Range from the Willamette Valley during an unexpected September snowstorm, losing only two along the way. By the next spring Townsend and business partner Harry Ogden managed to have their factory up and running. Perhaps Townsend's smartest move was hiring expert cheesemaker Peter McIntosh away from the factory at Woodland, Washington, to run his operation. McIntosh brought sorely needed expertise to a region in search of a sustainable commercial livelihood, and ended up setting the standard for cheesemaking in Tillamook. Townsend went on to operate ten creameries and cheese factories in Oregon and Washington and his White Clover brand butter became well known throughout the Pacific Northwest during the early twentieth century.

Until the mining boom of the 1860s, the future state of Idaho was a fairly barren, arid region many on the Oregon Trail simply passed through on their way to the much greener pastures of western Oregon or western Washington. As a result, the development of major industries of all sorts, including dairying and cheesemaking, lagged behind that of Oregon and Washington. Still, despite Idaho's considerably slower population growth and agricultural development, the state's reported cheese production grew rather dramatically between 1870 and 1880, shooting up from twenty thousand to over two hundred thousand pounds of cheese. A good portion of that production can be attributed to the Bear Lake Valley region located in the far southeast corner of the state adjacent to the Utah border. Mormon pioneers made their way north from Salt Lake City to this remote valley in the Rocky Mountains,

Olympic Ridge Creamery & Cheese Factory, Burlington, Washington, 1898. Photo courtesy Skagit Valley Historical Museum.

located at an altitude of over 6,000 feet, and started permanent settlements in the region in the 1860s. Led by Charles Rich, a disciple of Brigham Young, the Mormon community of Paris, Idaho, organized several cooperative endeavors including a sawmill, a tannery, retail shops, and several dairies. The cooperative system was a reflection of their religious principles: "[I]t is the purpose of the Lord to build up the poor," said Charles Rich, "if twenty or thirty persons engage in business and we make them more wealthy and others poor, what difference is there between us and the gentile world?" The dairy cooperative was located in the Noonan Valley near Paris and settlers choosing to join sent milk cows to the dairy in exchange for stock. The operation initially produced butter but later added swiss-style cheese to its product lineup. Because it was less laborious to produce, the cooperative eventually converted its production to cheddar. By 1883 the operation was using the milk of over fifteen hundred head of cattle to make two hundred thousand pounds of cheese annually.

The transition between settlement and the emergence of a dairy products industry in the Pacific Northwest was swift, the speed of this transformation due in part to the fact that the region's dairy farms evolved in catch-up mode. As far back as the eighteenth century, farmers in Connecticut, Massachusetts, and Rhode Island had been making and shipping hundreds of thousands of pounds of cheese to the developing southern colonies and to the West Indies. By the nineteenth century the state of New York was coming into its own as the dominant cheese-producing area in the nation, and in 1851, Jesse Williams started what is generally considered to be the first formal cheese factory in the United States in Rome, New York. A shift between farmstead cheese-making, in which farmers made cheese using the milk of animals kept on their own farms, to factory cheesemaking, in which milk from multiple farms was aggregated and processed at a dedicated facility, was already well underway in the United States by the time settlers were traveling over the Oregon Trail toward the Pacific Ocean. In 1890, total national cheese production already exceeded two hundred fifty million pounds.

The Pacific Northwest dairy industry, which came to life temporarily during the Hudson's Bay Company era, was propelled fully into being by the thousands of settlers who came to the region later in the nineteenth century. Many brought dairying and factory cheesemaking experience with them, and the rest brought a taste for cheese and other dairy products they'd become accustomed to back home. In addition, entrepreneurs from the East and Midwest with dairy industry knowledge and experience wasted no time taking advantage of the cheap land, plentiful labor, and endless green grass for grazing cattle, in the process speeding the regional cheese industry's development. Eventually, a growing resentment against the high cost of imported goods from the East stimulated production further, as the *Morning Oregonian* scolded the state's farmers for the "prodigious deficit between what we produce in the simple line of domestic foods and what we consume." Oregon, the paper said, had been forced to import all manner of food and goods from the East Coast including two hundred fifty thousand pounds of cheese in one year alone. The stage was set for the rapid expansion of the cheese industry in the Pacific Northwest into the twentieth century.

Adolf Sahli, circa 1915. Sahli operated the Stanfield Cheese Factory in Stanfield (Umatilla County), Oregon. Photo courtesy Oregon Historical Society.

Chapter 3

From Farm to Factory:
Cheese Becomes Big Business

By the dawn of the twentieth century the Pacific Northwest had been utterly transformed. No longer the remote, vaguely mythical wilderness that Meriwether Lewis and William Clark had explored, during the intervening decades the region had been mapped, carved up, and allocated by the various nations competing for dominion over it. The vast area once known as Oregon Country was now divided between the United States and Britain along the 49th parallel, and the portion belonging to the United States was further divided into the political entities of Oregon, Washington, and Idaho, now full-fledged states. The Northern Pacific Railroad had forged a connection between these newest states and the East Coast during the late 1880s, facilitating the speedy distribution of the region's lumber, coal, and all manner of manufactured goods to points east. By 1900 over one million residents inhabited the Pacific Northwest and just a decade later the region's population had more than doubled to 2.5 million. In Portland, the region's largest city, residences and commercial buildings had replaced the once dense forests adjacent to the Willamette River. Seattle was beginning to come into its own as a major urban center, fueled by the Klondike Gold Rush that transformed the city into a bustling commercial gateway to the Yukon Territory. Streetcars busily coursed the byways of both Portland and Seattle, delivering the growing population of riders to shops, boarding houses, manufacturing facilities, and, at least until Prohibition hit in 1920, taverns and saloons. The Pacific Northwest was positively teeming with the optimism of a new century.

Rapid Growth and Change

While residents of the Pacific Northwest were hard at work recreating the trappings of civilization that they had left behind, the nation as a whole was also growing rapidly. By 1900, the United States was a bustling commonwealth of forty-five states with a combined population of over seventy-six million people. The rise of large cities like New York City, Chicago, and Philadelphia, each with over a million residents, gave rise in turn to a system of large-scale agricultural production to meet the needs of a populace increasingly isolated from their food sources. Draft horses were replaced by steam-powered threshing machines. Seed drills mechanized the planting of fields. An increasingly complex railroad infrastructure provided a fast and efficient means of delivering farm produce to hungry cities. Rural, agrarian subsistence foodways of all kinds were rapidly evolving into large-scale food industries.

Science and technology propelled a rapid growth in production of cheese and other dairy products during the first decades of the twentieth century. Advances in the understanding of genetics that had begun with Gregor Mendel's observations of his garden peas in the 1860s led to a new age of bovine breeding. Twentieth-century dairy farmers could select the best cattle for milk production and, over time, a strong lineage of milking breeds including Shorthorns, Jerseys, and Holsteins were established across the Northwest. Farmers were able to get more milk, and thus more profit, from the same number of cows. Scientific innovations also changed the way milk was handled and processed. A simple milk test developed at the University of Wisconsin in 1890 by Professor Stephen Babcock revolutionized milk handling. The Babcock test enabled milk facilities and creameries to measure milk quality in the form of percentage of butterfat and pay farmers based on the results. Farmers and factories gained a powerful tool to analyze and improve the quality of their milk and dairy products.

Mechanical inventions transformed the formerly time- and labor-intensive tasks of making butter and cheese. The development of a mechanical process for separating cream from milk using centrifugal force during the late nineteenth century revolutionized a chore that previously could only be accomplished by time and gravity. The cream separator and its successive improvements enabled the increasingly efficient and profitable production of butter on a larger scale than had previously been possible. Inventors debuted specialized cheese vats that used steam to heat milk for processing, replacing messy, not to mention dangerous, wood fires. Steam-powered curd mills alleviated the need for labor-intensive hand cranking. Christian Hansen of Denmark developed a standardized rennet solution that allowed for more streamlined cheese production techniques, giving larger yields as well as better and more consistent quality. Newly designed cheese presses were capable of handling dozens of wheels of cheese at the same time, enabling producers to make cheese faster and more efficiently than ever before.

Across the nation, a new type of institution, the agricultural college, delivered the gospel of scientific progress and from it emerged a new generation of farmers and dairymen. Through the Morrill Act of 1862, the federal

government had granted federally owned land to the states for the establishment of institutions of higher learning devoted to science, agriculture, and
engineering. During the late nineteenth century three of these land-grant
institutions were established in the Pacific Northwest. Oregon Agricultural
College in Corvallis, Oregon, was the first to open to students in 1868, followed
by Washington State Agricultural College and School of Science in Pullman,
Washington, and the University of Idaho in Moscow, Idaho, both of which
first opened to students in 1892. All three institutions established academic
programs devoted specifically to agriculture and dairying as part of their mission. The need for these programs was more than obvious; as the *Spokane
Spokesman-Review* put it, "[t]he time of the haphazard go-as-you-please tiller
of the soil has passed. Agriculture is now a specialized science[.]" Each college
also established a working dairy farm and milk-processing facilities that offered
hands-on training as well as real-life laboratories for students and professors.

In addition to their academic programs, each of the Pacific Northwest's
agricultural colleges developed outreach programs directed toward local farmers. Congress provided funds for land-grant institutions all over the country
to establish Experiment Stations to conduct agricultural research and develop
the still-familiar Cooperative Extension network to benefit surrounding communities. Through another popular nationwide program called the Farmers
Institute, a developing body of scientific knowledge was disseminated to
farmers. Due to the long distances between the Northwest's three agricultural
colleges and the farming communities they served, each regional school took
to the rails in Farmers Institute demonstration trains, traveling classrooms
with lecture space and exhibits devoted to a variety of agricultural subjects.
College professors and guests lectured on a wide range of topics ranging from
diseases common to fruit trees, to draft horse selection, dairying, and building of rural homes. On one trip in 1909, Oregon Agricultural College's train,
known as the Demonstration Special, coursed through seven cities in eastern
Oregon bearing the empowering slogan "A crop for every acre, every year."
Washington State College got creative, in at least one case employing a demonstration boat to transport the series to farmers in Puget Sound's San Juan
Islands. Thousands of dairy farmers, creamery managers, buttermakers, and
cheesemakers passed through the variety of courses, degree programs, and
outreach seminars offered by all three regional land-grant institutions over
the early decades of the twentieth century.

In 1900, Assistant Professor F. L. Kent of Oregon Agricultural College set out
on a trip to assess the state of Oregon's dairy industry. The trip was notable in
part because Kent traveled throughout the dairy country of western Oregon

entirely by bicycle. An avid cyclist, Kent belonged to an Oregon Agricultural College faculty bicycling group called the "Century Club." On free weekends, the group cycled through the Willamette Valley on one-hundred-plus-mile rides. So enthusiastic were these cyclists that four of them, including Kent, had bicycled from Corvallis to San Francisco (over five hundred miles each way on today's freeways) during the summer of 1896 to witness the commissioning of the battleship *Oregon.*

Aside from his avocation, Professor Kent represented a new breed of academic specialty – the dairy scientist. His 1900 report on his research trip details a rapidly expanding dairy products industry in the state. By his assessment, the "factory system of making butter and cheese in the state is a development of the past ten years." Kent recorded 84 factories producing dairy products operating in Western Oregon, 35 of those facilities producing cheese—and the numbers, by his own admission, were incomplete because he hadn't been able to visit the eastern part of the state. Kent estimated the total output of cheese factories in the state at just over one million pounds.

Cheese production grew steadily in all three Northwest states during the first decades of the twentieth century. By 1919 Oregon boasted 74 cheese factories which produced a remarkable 9.4 million pounds of cheese; that same year Washington produced 2.3 million pounds of cheese and Idaho 2.6 million pounds. The numbers show that the Pacific Northwest was coming into its own as a dairy production center; as early as 1909 Oregon was already one of the ten largest cheese-producing states in the nation.

One reason for Oregon's growing dairy production numbers was the rapid growth of the dairying region near Tillamook Bay on the Oregon coast. By the early 1900s Tillamook County was coming into its own as one of the state's dairy production centers, with fourteen creameries and cheese factories operating by Professor Kent's count. As was the case in dairying regions across the Pacific Northwest, Tillamook's dairy farmers started out making dairy products on their own farms. The earliest white settlers in the Tillamook area had kept cattle and made and sold butter since the 1850s, but one big hurdle to making cheese was that few knew how to do it very well. Civil War veteran Merriman Foland traveled with his wife to California after the war, then to Oregon, and eventually settled in the town of Beaver north of Tillamook in 1878. While in California, both Foland and his wife had worked in a cheese plant, and the experience emboldened him to try his hand at the craft. Foland's earliest attempts at making cheese were said to have turned out quite badly, so badly in fact that some wheels of cheese swelled up (due to gases produced by internal bacteria) and rolled right off the shelves of their own accord. But Foland kept at it and eventually was able to produce a consistently good quality cheese.

Beaver Cheese Factory, Beaver, Oregon, 1893. The Beaver Cheese Factory was one of Tillamook County's first cheesemaking cooperatives. Pictured are Ike Heiner, cheesemaker, and Elizabeth Foland. Photo courtesy Oregon Historical Society

Neighbors like Jasper Smith, who later became a state representative, followed suit and developed their own farmstead cheesemaking operations.

The long distance between Tillamook and market centers in bigger cities like Portland and the significant geographical barrier presented by the Coast Range mountains made land transport of its products to market arduous, at least until the arrival of the railroads. The multiple hardships of living in the remote region created huge barriers to individual success; by the turn of the century a number of farmers in the area formed manufacturing cooperatives and built facilities that processed the milk from member farms into butter and/or cheese; among these were the Maple Leaf Creamery and the Tillamook City Creamery. Because maintaining farms, milking cows, and producing dairy products took up the vast majority of their time, several of the operations hired Carl Haberlach as a broker to market their products. Haberlach, a lawyer by training, stumbled into the cheese business when he took a side job filling in temporarily for manager George Williams at Tillamook City Creamery in 1903. The brokering arrangement was simple: Haberlach was paid a $15 monthly salary, but he made real money buying cheese and then selling it to the wholesale trade for as much as he could, pocketing the profits. Haberlach proved to have a particular talent for sales and soon other local cheese factories were eager to avail themselves of his expertise. By 1908 Haberlach was acting as the sales representative for thirteen Tillamook area factories.

Tillamook Creamery, Tillamook, Oregon, 1915. Photo courtesy Oregon Historical Society, Jeff Uecker Collection.

While Carl Haberlach proved very successful at selling cheese, returns varied from facility to facility. Each plant's product looked and tasted different, and even within an individual factory the cheese produced was not always consistent from month to month or year to year. To further complicate matters, independent-minded area cheesemakers had varying opinions about the best manufacturing practices; some are said to have followed a so-called "sweet cheddar" process introduced from Wisconsin while others followed a cheddar recipe introduced by Peter McIntosh. Facilities and equipment were clean and new at some factories and older and/or improvised at others. The resulting inconsistencies led to lower prices for everyone.

After several failed attempts at working cooperatively, in 1909 several (sources vary on the precise number) of Tillamook County's cheese factories formed the Tillamook County Creamery Association (TCCA). In its initial stages the cooperative resembled something like a fraternal association with the following stated purpose:

[T]o bring the patrons of the different creameries in Tillamook County Together, establish and maintain just and cordial relations among them and by cooperation to advance their common interest; to foster and encourage domestic and foreign trade pertaining to the interest of Tillamook County. To acquire and preserve and

disseminate valuable business information; and to adjust contro-
versies and misunderstandings between its members and generally
secure to its members the benefits of cooperation in furtherance of
their legitimate pursuits.

In this earliest phase the TCCA's primary focus was standardization.
Following emerging industry trends, the TCCA developed collective quality
standards that it mandated for every member factory. Toward that end they
hired Fred Christensen, then a cheesemaker at the area's Cloverdale plant,
as its chief inspector. One of Christensen's first tasks reportedly was weed-
ing out the heavy drinkers among the cheesemakers and substituting sober
ones—a move that no doubt had an immediate positive effect on the quality of the region's cheese. TCCA also developed another important quality
control measure, forming a cow-testing association to ensure that farmers
maintained healthy herds of cattle and received advice and assistance in their
care and management. The end result would be, it was hoped, greater returns
for everyone. Sales were still negotiated individually, however, between each
individual factory and Carl Haberlach.

Cheese production and sales increased in Tillamook nearly every year after the association was formed, no doubt assisted by the arrival of the first rail
line in Tillamook in 1911, constructed by the Pacific Railway and Navigation
Company. By 1917 the TCCA had evolved into a full-fledged marketing cooperative that not only enforced quality standards among its member factories
but also sold its members' products collectively. Efforts toward branding began
in earnest and the cooperative began advertising using one name, Tillamook,
in 1918, thanks to the efforts of Dave Botsford, who came to Tillamook from
a position as sales manager at Portland department store Meier & Frank. Area
cheesemakers had often imprinted the rinds of their cheeses with the name of
their creamery; soon they were printing the word "Tillamook" onto the rinds
of their cheeses, an effect they achieved by rolling the wheel of cheese over an
inked wooden plank. The TCCA subsequently developed an ad campaign that
advised consumers to look for the cheese with "Tillamook on the Rind." The
results speak for themselves: TCCA production more than doubled between
1908 and 1918, reaching over five million pounds annually.

While Tillamook's dairy farmers prospered, farmers farther south on the
Oregon coast in Coos and Curry counties were not far behind. Among the earliest cheesemakers in the area were the Rogers brothers, Stephen and Anson,
who, along with their brother Amos, had come to southern Oregon all the
way from Danby, Vermont. Both Anson and Stephen Rogers established dairy
farms on the South Coos River, where they produced prodigious amounts of

butter and cheese. Anson Rogers' wife, Elizabeth, assisted in making cheese, a skill she had acquired in her native Vermont; each wheel of the family's cheese was said to have weighed thirty pounds. Neighbors H. E. and Everett Bessey, also brothers, kept cows and made butter and cheese as well. Swiss-born Fred Moser learned the dairy and cheese business from his father and operated his own creamery in Switzerland as well as one in Italy before immigrating to the United States in the early 1890s. After surveying the conditions for farming in Idaho and California, Moser settled on the southern Oregon coast; he started a cheese factory on a forty-acre farm along the Coquille River in Gravelford and later opened a second operation, the North Fork Cheese Factory, in nearby Lee. Captain W. C. Harris of the Sumner Creamery began making cheese in 1901 with the assistance of cheesemaker William Blackmore, a recent graduate of Washington State College. Harris found quick success with his product; by the next year his cheese had won a blue ribbon at the Oregon State Fair and Harris was entertaining offers from San Francisco cheese brokers. In 1910 Harris switched from making cheese to bottling whole milk in his state-of-the-art dairy parlor.

The single largest dairy products factory of the turn of the century in Coos County was the Coos Bay Mutual Creamery. Formed in 1892 as a farmers' cooperative, it produced over one hundred fifty thousand pounds of butter and fifty thousand pounds of cheese in 1899. By then many of the area's independent dairy farmers, including the Rogers family, had begun sending their milk to the creamery rather than produce dairy products on their farms themselves. The profusion of streams and tributaries in the region enabled dairy farmers easy river access to this and other of the region's cooperative creameries. At one time as many as one hundred boats were engaged in the daily business of transporting milk from cans placed daily on farmers' docks to the butter and cheese factories downriver.

The southern Oregon coast's growing reputation as a dairying center drew the attention of East Coast entrepreneurs. In 1914, Paolo Griffon came to Oregon with a reported (and probably exaggerated) one million dollars and the idea of making sheep's milk cheese in the state. Griffon, a representative of an unnamed Italian corporation based in New York, toured areas of Douglas and Coos counties in search of a large tract of land, variously reported as fifty thousand or five hundred thousand acres, on which his company hoped to develop a sheep ranch and develop an associated dairy and cheesemaking operation. According to news reports Griffon actually purchased 480 acres south of Coos Bay in the Beaver Hill area in August of 1914. Despite the giddy excitement these developments generated in the local press, nothing came of these early efforts toward establishing an Italian-owned sheep's milk cheese

Coos Bay Mutual Creamery, Marshfield, Oregon, circa 1910. Photo courtesy Coos Historical and Maritime Museum, gift of Jack Slattery.

factory in Oregon. Italy entered World War I just months later in May 1915, and the Italian company most likely scrapped its West Coast expansion plans.

While the Coos Bay region possesses similar agricultural advantages to those of its neighbor Tillamook to the north—a relatively mild climate and proximity to ports—the region's farmers did not gravitate to the idea that dairying was their ticket to collective prosperity as readily as Tillamook farmers did, in large part because the timber industry dominated the south coast region's economy. But some saw its dairying potential. "Tillamook cheese is known throughout the Northwest because of the standardizing of the name," remarked Professor Hector McPherson of Oregon Agricultural College; "there is no reason in the world Coos County cannot do the same. It means a higher price for the product." Over the course of the next few years this perceived negligence began to change; perhaps the emerging success of Tillamook's cooperation experiment helped make plain to dairy farmers across the state the potential benefits of acting together toward a common economic purpose. In 1916, thirty-three cheese factories organized to form the Coos and Curry Counties' Cheesemaker Association, a marketing cooperative to market their collective output of cheese. Interestingly enough, just a year earlier the Tillamook County Creamery Association had sent representatives to investigate conditions on the southern Oregon coast and the southern Oregon farmers subsequently sent representatives to Tillamook in a sort of mutual courtship dance. News stories of the day hinted at the possibility of the Coos and Curry group uniting or at least working with the TCCA to the mutual benefit of

Workers pack freshly churned butter at the Brown Farm creamery, Nisqually, Washington, 1913. Photo courtesy Washington State Historical Society.

both regions, but a united coastal cooperative organization never came to pass. Like the Tillamook County Creamery Association, the Coos-Curry cooperative went on to develop quality control mechanisms and testing standards in order to ensure a more uniform and thus more marketable product across its membership.

Given the remoteness of Oregon's coastal regions, particularly before railroad transport reached these areas, it is somewhat surprising that Tillamook and Coos counties were early leaders in state dairy and cheese production. But farmers in coastal regions had one big advantage over landlocked farmers in other parts of the state—access to ocean transportation. Both Tillamook to the north and Coos Bay to the south developed a lively trade shipping all manner of products, including lumber and dairy products, to markets in San Francisco, Portland, Seattle, and even Asia. Newspapers in Marshfield (the city became known as Coos Bay in 1944) regularly reported the comings and goings of ships laden with locally produced goods, regularly citing the farm produce, butter, and cheese on board. "The *Redondo* sailed this morning for San Francisco," read one article; "she had a big cargo of lumber and some miscellaneous freight, including about 100 cases of cheese." Another headline trumpeted, "*Iaqua* Sails South Sunday: Steamer Leaves With Large List and Good Cargo, Butter and Cheese Shipped." Still, all of that reliance on river

Brown Farm

Alson "A. L." Brown was a Seattle attorney and the son of a prominent Seattle lumberman and landowner, Amos Brown, one of the city's earliest citizens. After his father's death in 1899, Alson Brown decided to become a farmer rather than a self-described "bum lawyer," and purchased over 2,000 acres of land near Olympia, Washington. The area, once occupied by the Hudson's Bay Company and its Puget Sound Agricultural Company satellite at Fort Nisqually, consisted in large part of marshy wetlands adjacent to Puget Sound that were not terribly conducive to cultivation. Undeterred and with money to burn, Brown solved this problem by engineering an elaborate four-mile long system of dikes that held back the waters of Puget Sound and allowed him to develop an extensive farming operation.

By 1914 Brown had grown his operation into a diverse mega-farm that, at its peak, was said to have housed upwards of 4,000 chickens, 1,000 turkeys, 1,200 pigs, 300 dairy cattle. The farm's primary dairy products were fluid milk and butter, but the on-site creamery also made buttermilk cheese, a simple fresh cheese made from buttermilk, the by-product of butter production. Brown Farm generated a dizzying variety of farm-produced goods, including eggs, poultry, fresh cuts of beef and pork and at least ten varieties of sausage. The operation required a considerable crew of as many as 80 men to keep the operation running smoothly, working both day and night shifts.

One of the hallmarks of the Brown Farm operation was its emphasis on direct farm-to-consumer production and distribution. Yet despite the ambitious vision, Brown declared bankruptcy in 1918, and the farm's animals and equipment were auctioned off over a period of several years. Reports vary on the reasons for the decline, but Brown's plan to ship products to individual consumers was stymied by the disruption of regular train service when the U. S. Army established Camp Lewis (later Fort Lewis) near the farm in 1917. Ultimately, Brown's experiment demonstrated that even a well-financed, large-scale farmstead operation would have considerable difficulty competing in the factory era.

The Brown Farm property changed hands multiple times over the next several decades and eventually became the Nisqually National Wildlife Refuge in 1974. A. L. Brown's dike system, by then over a century old, was removed as part of an extensive restoration effort in 2009.

and ocean transportation also had a dark side; unpredictable currents and rough seas could take their toll and wrecks were not uncommon. The 1913 headline "Launch Sinks With Cheese" surely struck a chord in the Coos Bay area's port-dependent dairy community.

While Oregon cheese production thrived during the early years of the twentieth century, production in Washington was significantly lower. This was due at least in part to the rapid growth of the condensed milk industry, which dominated western Washington during this period; milk that might have otherwise gone to making cheese went instead to the state's rapidly expanding condensary market. By 1910 Washington was home to thirteen condensed milk factories compared to just six in Oregon. The Auburn Creamery Co., which operated a butter factory in Auburn as well as a cheese factory in nearby Kent, was one of the largest and most successful cheese producers in King County, Washington, during the early twentieth century. Proprietor A. H. Meade marketed "Meade's Gold Medal" brand cheese, which was reportedly good enough to have brought "fancy prices since the brand was established." Operating in the area was a challenge, however, because the Pacific Coast Condensed Milk plant in Kent competed for milk from area dairy farmers and often paid better prices; Meade's cheese plant was compelled to close temporarily in 1901 due to lack of milk. The Pacific Coast Condensed Milk Company's Carnation Evaporated Milk went on to become a nationally known and distributed brand and was eventually acquired by Switzerland-based Nestlé in 1985.

Because of the active condensary industry on the west side, most of the cheese and butter produced in Washington came from the eastern side of the

Cheney Cheese Factory, Cheney, Washington, 1906. Cheesemaker Frederick Reuter is pictured at right.

Photo courtesy Cheney Historical Museum.

state. Eastern Washington's largest cheese factory was the Cheney Creamery, started by Francis M. Martin and business partner Willis Hubbard in Cheney, just southeast of Spokane, in 1890. By 1897, the facility and its satellite plants in the nearby towns of Tyler and Medical Lake collectively produced enough cheese to reign for a period of time as the largest producer in the state of Washington. A good portion of Martin and Hubbard's success was likely due to the fact that they hired German immigrant Frederick Reuter to manage their cheesemaking operations; Reuter had a dairy diploma from the University of Wisconsin and had served as a dairy instructor both there and at the University of Idaho in Moscow. Martin and Hubbard eventually sold out to the fast-growing Spokane-based Hazelwood Creamery in 1899. Martin seemed to have a golden touch for business; he went on to found the F. M. Martin Grain and Milling Company, which he sold to Nabisco in the 1940s. Frederick Reuter started his own cheese factory, the Cheney Cheese Factory, in 1903 but was forced to give it up several years later due to illness.

Hazelwood was one of the Inland Empire's big dairy success stories of the period. David and George Brown's father had operated the original Hazelwood farm in Galena, Illinois, where he bred shorthorn cattle. The Brown brothers were drawn to Spokane, Washington, by a cattle breeder who visited their family's farm in Illinois; they eventually succumbed as many had before them to the utopian dream of going west. Along with another Illinois native, John L. Smith, the Browns established their own Hazelwood Farm and started a dairy and butter creamery operation near Spokane in 1892. The business grew quickly with increasing demand, and Hazelwood proceeded to acquire and establish a number of smaller creamery and cheesemaking operations in eastern Washington and even opened a milk processing plant in Sioux City, Iowa, in 1899, when it couldn't secure enough milk in the Pacific Northwest to support its growing operations.

The Smith-Brown partnership turned Hazelwood into a multi-tiered commercial enterprise, branching into wholesale operations as well as cold-storage services. Hazelwood even dabbled in the retail arena, in 1900 establishing the Hazelwood Cream Co. in Portland, which initially sold butter and ice cream but later branched out into ancillary products such as candy. The Hazelwood Cream Store attracted customers with events including musical performances and the company's Portland operations eventually grew into a string of several restaurants and a retail outpost that sold candy and ice cream at popular tourist attraction Multnomah Falls in the Columbia River Gorge. In its food-service capacity Hazelwood became a major purchaser of dairy products in the state, a fact that stirred up some tension in Tillamook since Hazelwood purchased most of the output from two cheese factories that had not joined the Tillamook County Creamery Association, the Long Prairie and Red Clover factories. In 1926, Hazelwood's retail operations were purchased by Seattle-based Western Dairy Products Company, an ambitious, fast-growing dairy corporation that owned and operated fifteen dairy plants (mostly ice cream producers) in Washington and Oregon, as well as California Dairies, Inc. in Southern California. Western continued using the Hazelwood name for a period of time and the Hazelwood restaurants continued to operate in Portland through the 1920s but the once large and influential dairy corporation eventually faded into history.

Another important dairy center in eastern Washington was located in the central part of the state, nestled up against the east side of the Cascade Mountains. The Kittitas Valley grew into a cattle ranching and grazing center in the nineteenth century; the area's proximity to the Puget Sound area just on the other side of the mountains made it a convenient repository for cattle that could be grazed and fattened on the flat, grassy east side and transported back west over the Snoqualmie Trail to Seattle to be sold. The area's population grew considerably after coal was discovered at Roslyn in 1883 and after the Northern Pacific Railroad made its first stop at the Ellensburg depot in 1886.

The dairy story of the Kittitas Valley is one of successive mergers and consolidations, many involving prominent businessman Briggs F. Reed. Reed, an Illinois native, operated a variety of businesses including a stock ranch and the Okanogan Stage Coach Company before starting the Ellensburg Creamery Company in 1894. B. F. Reed worked multiple sides of the dairy business: in addition to owning the creamery he raised and sold cattle and also had a hand in Miller Reed Pease, a wholesale distributor of dairy products located in Seattle's merchants' row on Western Avenue. Miller Reed Pease operated four dairy products factories in the state (including Reed's own factory in Ellensburg) and was known for its Jersey Lily brand butter, which was distributed in the Seattle area. Miller Read Pease later became Turner and Pease, and the wholesale distribution company is still operating as a butter distributor in Seattle.

Around the turn of the twentieth century as many as seven dairy plants in addition to Reed's were operating in the Ellensburg area, most of which produced butter. In 1900, Simon Whipple, one of Reed's former partners, started the Kittitas Creamery Company. The next year that company acquired the Cloverdale Creamery of nearby Thorp, started in 1893 by cheesemaker John Goodwin. Goodwin was said to have suffered a serious injury while haying in 1898, a factor that may have contributed to the sale of his business. The Alberta Creamery Association opened its doors in 1900, producing butter

Merchant's Row on Western Avenue in Seattle, Washington, 1905. Photo courtesy Museum of History and Industry, Seattle, PEMCO Webster & Stevens Collection.

that it shipped primarily to the Puget Sound region. In 1914 Alberta Creamery combined with the Spring Creek Creamery, established by J. P. Sharp, to form the Consolidated Creamery Company. By then, serial entrepreneur Reed had gotten out of the dairy business altogether and into milling and selling grain instead at his Ellensburg Milling Company.

Regardless of who owned what and when, little of the commercial prosperity east of the Cascade Mountains would have been possible without one of the most significant agricultural developments ever to arrive in the Pacific Northwest—irrigation. While the western halves of Washington and Oregon, along with northern Idaho, are blessed by generally plentiful rainfall, fertile soil, and a mild climate, points east of the Cascade Mountains as well as in southern Idaho possess to varying degrees a less hospitable high desert climate with precipitation in some areas averaging as little as eight to ten inches annually. As white settlers began to populate areas east of the Cascades because it was cheaper and less crowded than more desirable areas to the west, they appropriated precious water wherever they could find it. Early missionaries—like the Whitmans at Waiilatpu in what would become the eastern part of the state of Washington and the Spauldings at Lapwai in latter-day northern Idaho—had diverted water from nearby streams as a matter of course to support their struggling agricultural operations and later settlers faced the same challenges. Those who flocked to the region to work in the gold and precious metals mines that sprang up during the late nineteenth and early twentieth century remained because irrigation made it possible for them to stay and forge a sustainable livelihood.

Eventually larger-scale irrigation projects came to the region, some privately funded and others developed and administered by the United States Reclamation Service, formally established by the federal government for the express purpose of encouraging settlement of government-owned land in the West. The Carey Act of 1894 and its successor, the National Reclamation Act of 1902, functioned much like the Homestead Act of decades past, attracting new waves of settlers with the prospect of cheap albeit arid land along with the water and/or the financing to irrigate it. Over the early decades of the twentieth century dozens of dams were constructed across the more arid parts of the Pacific Northwest, diverting water from sources including the Snake River in Idaho, the Klamath River in southern Oregon, and the Yakima River in central Washington. Idaho's population more than doubled between 1900 and 1910, largely as a result of irrigation-related development along the Snake River, which courses through the southern part of the state. The once dry, sagebrush-strewn landscape in this arid section of the Pacific Northwest literally blossomed as settlers moved in and established productive farms of all kinds.

While water did not by itself create an environment for dairying, the alfalfa that water nurtured did. Alfalfa deposited nitrogen in the soil, making it more fertile for the area's cash crops like sugar beets and potatoes and did double duty as a cash crop of its own when used as fodder for cattle and sheep. As dairy farmer W. H. Carpenter of the North Yakima Valley in Washington discovered, however, it was a significant challenge to convince local farmers to feed their alfalfa to cows and then sell him milk to make cheese rather than simply grow alfalfa and ship the hay out of state to make a quick buck. While the farmers complained to him that keeping cows required significantly more labor, Carpenter's reply was succinct: "What matters it if we farmers pay out money for labor when the products of that labor bring us many times as much as we pay out?"

Convincing area farmers to look beyond the immediate toward a distant but potentially prosperous dairy-based economy proved difficult. Carpenter recognized that a good part of his task in operating a local cheese factory would be farmer education; he noted poignantly that "my establishment is a missionary affair to some extent." He must have convinced at least a few farmers, as during the earliest weeks of business in 1894 his cheese factory brought

Paul Butter and Cheese Factory Paul, Idaho. Photo courtesy Special Collections and Archives, University of Idaho Library, Moscow, Idaho.

in an estimated sixteen hundred pounds of milk per day. Success seems to have been short lived, however. By the next year the local paper reported that Carpenter had begun to sell groceries at the creamery, likely in an attempt to diversify. Carpenter later took on a Mr. Bartlett as a business partner, and eventually cashed out entirely. Bartlett sold the operation in 1898.

Idaho's turn-of-the-twentieth-century dairy farms and cheese factories followed the irrigation projects that multiplied along the Snake River as it courses its way through the length of southern Idaho. The Minidoka Project east of Pocatello opened in 1909, when the first of several dams began generating power and diverting water to area farms. The number of dairy cattle in the area increased from less than two thousand in 1911 to seven thousand three hundred five years later. A number of creameries sprang up in the area to handle the growing milk supply and the first area cheese factory was established in 1915 when a group of farmers banded together, hired a cheesemaker, and started the Acequia Dairy and Produce Company. The factory made over nine thousand pounds of cheese in June of 1916, though it struggled to compete for milk from area dairy farmers who were sending it to the established butter creameries. A cheese factory also opened in nearby Paul, Idaho though it, too, struggled to obtain milk and eventually closed.

Sixty miles or so to the west in Twin Falls, the Milner Dam project brought similar prosperity to that area. In fact, so strong was the pull of agricultural promise that several dairy families from Tillamook, Oregon, including the Maxwell, Carlson, and Kunze families, migrated eastward to the Twin Falls area. According to Joseph Maxwell, "[we] were sold on the possibilities [in Idaho]." Gustav Kunze, well known as a cheesemaker in Tillamook, started a cheese factory in Buhl, Idaho, in 1912 called the Clover Leaf Factory (perhaps a wry reference to the Clover Leaf Factory in Tillamook). He operated the factory for several years before selling it to the Sandemeyer family just after World War I. Kunze went on to become a spokesman for the state's dairy interests as president of the Idaho Dairymen's Association, and later became president of the Twin Falls County Dairymen's Association, a dairy cooperative. Other outsiders such as John Kaeser, who owned a cheese wholesale business in Monroe, Wisconsin, called Badger Cheese Co., also recognized the region's potential; Kaeser opened a cheese factory in 1914 in Filer, Idaho near Twin Falls.

The Mormon settlements in the Bear Lake Valley of eastern Idaho would not have been sustainable without water. Mormon settlers constructed canals to siphon water from the Bear River and its tributaries, and the Bureau of Reclamation later completed a reservoir and a dam there. While the colony's early cooperatives, including the cheesemaking cooperative, dissolved by the early 1880s, dairying continued as a thriving industry in the region. By one account there were as many as seventeen cheese factories in Bear Lake County by 1897.

Among the most prolific of the area's dairy and cheesemaking families was the extended Kunz family, who brought their skills to the valley from their native Switzerland. Several family members were converted by Mormon missionaries in Switzerland and emigrated to the United States during the 1870s. Johannes Kunz III brought over two large copper kettles when he traveled to America, transporting them on board ship and rail to Utah, where his father, Johannes Kunz II, was already operating a cheese factory. The family established the Kunz Brothers Dairy in Bern, Idaho, in the Bear Lake Valley, though they operated separately from the church-run dairy cooperative. By 1885 the family was making four hundred pounds of cheese per day from the milk of three hundred cows. Johannes II's brother Samuel Kunz also established a series of cheese factories in the area, including one at Peagram and a later one at Bates in the Teton Valley near present-day Driggs, Idaho.

During the 1890s, Johannes Kunz III purchased land about fifty miles north of Bern in Williamsburg, Idaho, on the Blackfoot River. He and his sons William and John operated three dairies on the property, known within the family as the upper dairy (owned by the father), the middle dairy (owned by son William), and the lower dairy (owned by brother John). During the spring, the extended Kunz family would travel from Bern, Idaho, where they lived during the winter, north to the farms at Williamsburg. On the way the family gathered cows from various farms across the region and drove the growing herd northward, amassing a herd of several hundred along the way. The cattle drive took several days. All summer, the family and hired help milked the cows and made cheese. In the fall, the family would travel by wagon to regional markets at Soda Springs, Idaho, and throughout Utah and Wyoming to sell the cheese, where it was reportedly quite popular. At the end of the season the Kunz clan returned cattle back to each family that had provided the animals, along with a percentage of the earnings based on the percentage of milk their cows had provided. The seasonal journey between winter and summer pastures, known as transhumance, mimicked the itinerant pasturing traditions of the Kunz family's native Switzerland. Various members of the Kunz family continued making cheese in Idaho's Bear Lake Valley through the 1930s.

The Evolving Marketplace

As farm production expanded across the Pacific Northwest, a system of wholesalers, distributors, and brokers developed that mirrored that already established on the East Coast. These outfits dominated the buying and selling of farm-produced products. Farmers typically sold products to wholesalers on consignment, either directly or via brokers who facilitated the connection, receiving a percentage of the proceeds when the products were sold. Another tier of small-time merchants purchased small lots either directly from farmers or from wholesalers and peddled the wares independently, often door to door. Commission merchants located on Western Avenue in Seattle, Front Avenue in Portland, Railroad Avenue in Spokane, and in other regional cities grew to wield great power in the buying and selling of all manner of goods in the Pacific Northwest.

A farmer's arrangement with a wholesale merchant was not without its risks, since wholesalers generally possessed a significant amount of power in dictating prices. For example, a dispute developed in 1901 between cheesemakers in Tillamook and a Portland area wholesaler over the commission the merchant charged farmers for Tillamook cheese. While the merchant apparently insisted on a one-and-one-half-cent commission, the cheesemakers would not accept any more than a one-cent margin, arguing that their product was superior to other cheeses on the market. The merchant countered, as one might expect, that "unless [I am] are given a margin with which I can do business I will import the best cheese from the East." Farmers counter-threatened that they would seek other markets for their product. The merchant, unnamed in the press, apparently followed through on the threat to import cheese from the East Coast, though the actions did not seem to cause irreparable harm to Tillamook area cheesemakers. A similar dispute erupted in 1914, when powerful Chicago-based Swift and Co. took manager Carl Haberlach to task for selling Tillamook cheese to Seattle wholesale merchants for a cheaper price than Swift was offering in the Portland markets.

In other cases, corrupt wholesalers and brokers cheated farmers out of their products for their own profit. Petty criminals were known to brazenly misrepresent themselves as employees of established commission houses, only to make off with valuable goods, never to be seen again. Other shady wholesalers fabricated stories about lost or destroyed shipments, or claimed that the farmer's products arrived spoiled or were of poor quality and thus worth less money. In Seattle, Western Avenue wholesalers were said to have dumped farm produce into Puget Sound, defrauding farmers and driving up the price for the (instantly more scarce) goods. B. F. Childs was one of many cheated out of payment for his products. Childs, the owner of the Star Cheese Factory in Brownsville, Oregon, sold eleven thousand pounds of cheese to a broker in 1900. In exchange, the broker gave Childs a draft for $1313.26, payable in ten days. Upon being notified by the bank that the draft's legitimacy was in question, Childs traveled to San Francisco and found the broker, who flat out refused to pay. Upon further investigation, Childs discovered that the man possessed no property of value so any attempt to sue him for the balance owed would be useless—and of course the cheese was nowhere to be found. Such were the perils of the turn-of-the-century cheese business.

Stark power inequalities between farmers and merchants, frequent and seemingly random price fluctuations, and widespread corruption (or at least the perception thereof) were among the factors that led to the development of the earliest formal public farmers markets in the Pacific Northwest. Tacoma's C Street Public Market first opened in 1891 and was eventually replaced by the larger, grander Crystal Palace Market in 1927. In Seattle, the Pike Place Market was founded in 1907. So popular was the idea of an independent public market that thousands showed up in downtown Seattle on opening day, only to leave empty handed due to the fact that there were so few farmers in attendance. Portland's Carroll Public Market opened in 1914 on Yamhill Street. This was the public market of James Beard's youth, with its kaleidoscopic array of tomatoes, odd-looking morel mushrooms, and captivating variety of international specialties—the market he called "an education in food" in his 1968 memoir. The later established Portland Public Market, housed in a large building situated along the waterfront, was opened during the 1930s. Boise's earliest public market opened in 1915; at its inception one entrepreneurial cheesemaker offered a variety of products for sale including homemade cottage cheese, which "was gone before nine o'clock, and her baked beans, pickled beets and home smoked bacon found a ready sale."

Though farmers markets were very popular and certainly succeeded in connecting farmers and consumers, they ultimately did not make a significant dent in the wholesale merchant trade. More important in the overall evolution of food merchandising in the United States was a parallel retail trend of the period. Just as Pacific Northwest citizens were furiously lobbying for public farmers markets, Clarence Saunders was busy back east in Tennessee developing a unique retail concept: a self-service store. The idea of going into a market to select one's own basket of goods and then pay at a central location in the front of the store was revolutionary enough that Saunders received a patent on the concept in 1917. Saunders went on to found the Piggly Wiggly grocery chain; companies such as the Great Atlantic and Pacific Tea Company (A&P) in New York and Safeway in southern California followed his popular model. By the 1920s these and other chains began to dominate the landscape for consumer food purchasing across the nation. As these stores grew in size and influence they began purchasing cheese and other dairy products, among other things, directly from manufacturers, saving themselves money in by-passing wholesalers altogether.

The Pacific Northwest received a big economic boost when the U.S. entered World War I in 1917, as the region contributed a vast amount of resources and labor to the war effort. Portland and nearby Vancouver, Washington, grew into centers for shipbuilding along the Columbia River. Airplanes, then a new invention, were produced at the newly formed Boeing Company, headquartered in Seattle. Planes of the era were made of wood and spruce was particularly favored by aircraft builders for its strength and light weight. The former Fort Vancouver site along the Columbia River in southwest Washington was transformed into an enormous lumber mill by the U.S. Army's Spruce Production Division, which produced in excess of one hundred and fifty million board feet of lumber on the site from spruce trees sourced from all over the Pacific Northwest. Idaho contributed lumber, minerals from its mines, and farm produce including grain and wool to the war effort.

Farm production soared across the United States during the war years as grain, meat, vegetables, and dairy products were exported to Europe to support the failing economy there as well as to feed U.S. troops. While there was no mandatory food rationing during the period, the United States Food Administration (USFA), led by future president Herbert Hoover, initiated a nationwide conservation campaign in which households, merchants, institutions, and local governments were all encouraged to cut back or "hooverize," the pop-culture term of the period for conservation. The USFA issued pamphlets with a rather complicated weekly schedule of wheatless and meatless meals it hoped citizens would observe; Saturdays brought the added burden of a specifically porkless meal. Restaurants got into the act by establishing meatless and wheatless menu days. Several of the Pacific Northwest's larger dairy producers attempted to take advantage of the war era's heavy emphasis on meat conservation. In a 1918 ad campaign, one of their earliest, Tillamook cheesemakers turned the purchase of their products patriotic, proclaiming: "Tillamook Cheese is Delicious—and it Saves Meat!" Kristoferson's Cottage Cheese, produced by a Seattle milk distributor, also played to the prevailing winds of conservation by spinning its product as a cost saver: "as a meat substitute, our superior cottage cheese cuts your bills in two."

Another important impact of World War I for the nation's cheese industry was that it interrupted the supply of cheesemaking rennet, at the time produced primarily in Europe. The resulting worldwide rennet shortage caused great concern in cheese industry states across the nation. In Oregon, farmers were asked to save the stomachs of their calves for use in making rennet as the pioneers had in decades past. Some cheesemakers considered switching to pepsin, a similar enzyme derived from pig stomachs, that had the added bonus of alleviating the need to slaughter valuable calves. Oregon Agricultural College conducted a study of the differences between rennet and pepsin, finding that pepsin could be a satisfactory substitute in the process of making cheese.

Tuberculosis and the Rise of Goat's Milk

Because milk is inherently perishable, until the nineteenth century it was primarily consumed in the form of cheese or butter, or a soured product such as clabbered milk. But by the 1830s, around the same time that the Hudson's Bay Company was expanding its network of trading outposts across the Pacific Northwest, a fledgling milk industry was starting to emerge in the eastern United States. This early industry was one that we would hardly recognize today. Imitating longstanding European practices, for-profit milk purveyors in New York and other East Coast cities kept barns full of milk cows adjacent to breweries and the cattle consumed the spent grain from the brewing process. The animals were very poorly cared for and were often so weak that they could not stand up and had to be hoisted up on a lift in order to be milked. The resulting product, aptly termed "swill milk," was said to be a disgusting thin, bluish grey fluid that contributed to the spread of disease, malnutrition, and a rise in infant mortality. Temperance reformers, already lobbying fiercely against the social degradation caused by alcohol, added milk reform to their broader cultural and social agenda.

The rapid growth in fluid milk consumption was problematic on another level as well. For centuries, tuberculosis killed millions of people all over the world. Known by many euphemisms, "consumption," often characterized as a romantic disease suffered by dreamy poets, permeated popular culture during the nineteenth century. Despite its reputation tuberculosis was no joke—as recently as 1900 tuberculosis was the leading cause of death in the United States. Until the late nineteenth century scientists and doctors had not been unable to understand, let alone conquer, the disease. That began to change as a growing understanding of microscopic organisms like bacteria and viruses ushered in a new perspective. What has since been termed the "germ theory of disease" revolutionized modern medicine, and physicians and scientists could finally effectively treat all manner of communicable diseases like tuberculosis, yellow fever, rabies, and malaria.

German physician Robert Koch was the first to isolate the bacteria that cause tuberculosis in 1882. Soon it became clear, however, that there were several strains of the disease including a human strain and a bovine strain, each of which produced different symptoms. Though Koch himself inexplicably denied that the bovine strain could be passed to humans, scientists eventually proved otherwise; historians now estimate that as many as 30 percent of human tuberculosis infections in the United States may have been caused by the bovine strain of the disease. The chilling realization that cattle could—and indeed probably had—spread tuberculosis through the nations' meat and milk supplies set into motion a massive nationwide effort to eliminate infected cattle. Starting in 1917, tuberculosis tests were administered to cattle in every state of the union and those animals that tested positive were destroyed. In just over twenty years, 3.8 million cattle were destroyed in the United States; not until 1941 did scientists declare the cattle population virtually tuberculosis-free.

Enter the goat. Goats have been established in the New World for centuries; the Spanish introduced goats and horses to South and Central America in the sixteenth century and residents of Jamestown and Plymouth colonies maintained herds of goats on the east coast of North America during the 1600s. Goats made their way to the Pacific Northwest on the ships of the many European explorers who scouted its coastline. We've already seen that the fur trappers at Fort Astoria and its later incarnation, Fort George, kept goats, and that the Hudson's Bay Company kept many goats at Fort Vancouver and its satellite farms and outposts. While the so-called "poor man's cow" never went away, goats carried on essentially under the radar in subsequent years, unremarkable relative to the much larger, more productive, and more profitable cow.

During the late nineteenth and early twentieth century, however, the popularity of goats began to grow as Northwest farmers seized on the idea of raising goats for their hides. In addition, mohair, a fabric made from woven Angora goat hair, skyrocketed in popularity. Angora goats, native to Turkey, were introduced to the United States in the nineteenth century and pioneers all over the developing western United States began raising them for profit. Farmers soon discovered that goats thrived on eating dense underbrush and employed them as low-maintenance land-clearing machines in the fast-growing Pacific Northwest. Angora goat farms sprouted up across Oregon, Washington, and Idaho; Polk County in Oregon's Willamette Valley grew during this period to become one of the centers of mohair production in the United States.

By the dawn of the twentieth century goat's milk was coming into favor. Milking breeds such as Saanens and Toggenburgs were imported into the United States from Europe and some of the earliest goat dairy herds in the United States were established in California during the early twentieth century. The American Milch Goat Record Association was established in 1905 to create a breed registry to document the lineage of the rapidly increasing number of specialty dairy goats in the United States. Goat's milk developed a reputation for its health-restoring properties, cemented by claims that it was capable of curing a wide range of ailments including allergies, eczema, epileptic fits, and weak bones. Goats also gained a popular, though undeserved, reputation as tuberculosis free, cementing the perceived connection between goat's milk and good health. The *Morning Oregonian* was moved enough by

Encaria brand bottled
goat's milk label.
Courtesy Bill Moomau.

the issue to raise the rhetorical question "Why Not Goat's Milk?" situating it as an appealing alternative both for farmers (as goats were smaller and thus required less care and feeding) and consumers (goat's milk was healthier, especially for babies).

Goat's milk became popular enough that by 1920 there were an estimated two hundred dairy goat farms across the Pacific Northwest, enough to support the founding of the Puget Sound Milch Goat Breeder's Association in Washington and the Oregon Goat Milk Association in Oregon. The *Seattle Daily Times* trumpeted the emerging Washington dairy goat industry, holding up Encaria Goat Farm of Woodinville, Washington, as one of the leaders. Encaria was owned by Melvin "M. P." Eggers. His herd of one hundred and ten Toggenburg goats was, according to the paper, the third largest goat farm in the nation, the largest being in southern California and the second largest at the Cook County Hospital in Chicago, where the milk from its thirty-six Toggenburg goats was dedicated to children in the tuberculosis ward.

Despite the proliferation of goat farms, M. P. Eggers described the market for goat's milk in the Northwest during this period as weak, at least partly because "most of the milk from the Seattle territory is shipped to California where demand is high. We hope to create a demand locally." Whatever they did, it must have worked, because Eggers began selling his Encaria brand goat's milk in downtown Seattle in the spring of 1921. A number of ads appeared in the *Seattle Daily Times* during the 1920s promoting milk from another local goat dairy, Sunnydale Goat Farm, owned by Martha and Owen Williams of present-day Burien (then known as Sunnydale). These higher-profile commercial operations were more the exception than the rule, however, and most goat milk producers in the Northwest remained small and relatively unnoticed. Still, the proliferation of classified ads offering goat's milk for sale in newspapers from Seattle to Portland to Twin Falls, Idaho, during the early decades of the twentieth century demonstrates the increasing popularity of goat's milk as an alternative to cow's milk. By 1928, Meyenberg Laboratories of southern California was producing canned evaporated goat's milk, which developed a significant national following. It would not be long before farmers at the many goat's milk dairies in the Pacific Northwest would begin making and selling cheese.

Though tumultuous economically and politically, the early decades of the twentieth century were largely prosperous in the Pacific Northwest. The region's economy expanded rapidly as settlers continued to swarm into the region and the area's vast stores of lumber, coal, and other natural resources generated continuing wealth and prosperity for the ever-growing population.

An increasingly developed railroad infrastructure facilitated the export of raw materials and manufactured goods to markets across the nation and the world, as did the opening of the Panama Canal in 1914. The number of cheese factories in Oregon, Washington, and Idaho multiplied and the better ones developed a reputation for their quality dairy products. By the 1920s the brand name Tillamook had become synonymous up and down the West Coast with good quality domestically produced cheese.

Nationwide economic expansion had transformed the dairy business, and rearranged the dairy farmer's role in relation to the marketplace. By the early twentieth century, most Pacific Northwest dairy farmers, like their counterparts across the nation, now simply milked cows or goats and sold milk or cream to a processing plant. The local factory, equipped with the latest production technology, trained employees, and state of the art milk-testing equipment, bottled and resold farmers' milk or manufactured dairy products like ice cream, butter, and cheese in larger quantities than all but the wealthiest farmers could ever imagine. The industry grew increasingly segmented, with milk production separated from dairy products manufacturing separated from marketing and distribution. On-farm production of products like butter and cheese grew rare as the rapidly expanding economies of scale made individual production financially out of reach.

Dr. N. S. Golding (pictured) of Washington State University in Pullman, Washington, developed the recipe for Cougar Gold Cheese in the 1930s. WSU still produces the cheese today. Photo courtesy of Washington State University.

Chapter 4
Expansion and Innovation

The World War I economic boom ended with the 1918 Armistice, sending the war-prosperous Pacific Northwest production engine into an economic slide. Thousands of lost jobs in the timber, shipbuilding, and aircraft industries contributed to a precipitous rise in unemployment and many who had migrated to the region for its wartime industry jobs simply left for greener pastures. The regional employment vacuum created an enormous problem for thousands of soldiers returning from wartime service and looking for work. Oregon officials held a Reconstruction Convention in 1919 to address the state's future in the face of feared postwar economic and social upheaval. Oregon Governor Withycombe, incidentally the former head of the Experiment Station at Oregon Agricultural College, proposed in his biennial address that the state purchase private lumber lands, subdivide them, and sell tracts to returning war veterans. The state of Washington held its own Reconstruction Congress, which recommended a wide-ranging slate of programs including the development of hydroelectric power to stimulate the economy.

The postwar years also ushered in what then-Secretary of Agriculture Henry Wallace ominously termed an "agricultural depression" in the United States. Previously astronomic export levels of agricultural commodities plunged, sending farm production into a tailspin. High wartime prices plummeted and crop production dropped as a result. Farmers who had expanded to accommodate (and profit from) increased wartime demand found themselves severely overextended during peacetime and many lost their farms. Severe drought conditions during 1918-1920 only exacerbated problems for farmers in the Pacific Northwest. Nationally, cheese production dropped 12 percent between 1919 and 1920, and did not reach 1919 levels again until nearly a decade later. In the Pacific Northwest, the previously rapid pace of growth in cheese production slowed considerably. Tillamook production levels increased 10 percent from 1919 to 1921, but dropped slightly in 1922 and remained essentially level until the mid-1930s.

Postwar economic instability contributed to the rapid growth of the agricultural cooperative movement in the United States. Though by no means a new concept, farmer-owned agricultural cooperatives became increasingly common during this period and were organized by wheat and fruit growers, egg and poultry producers, and cotton farmers across the nation. At the peak of the cooperative movement in the United Sates during the 1920s there were as many as fourteen thousand agricultural cooperatives across the country, over two thousand of which were dairy cooperatives. Congress enabled the cooperative movement's expansion by enacting the Capper-Volstad Act of 1922, which protected cooperatives against accusations of price fixing and anti-trust litigation. Perhaps one of the more appealing aspects of cooperative organization was that, until 1951, agricultural cooperatives were treated as tax-exempt entities by the Internal Revenue Service.

Though on its face the term cooperative suggests a communal system of sharing work and profit (in fact, the whiff of socialism implied by the name made cooperatives suspect to the conspiracy-minded of the period), agricultural cooperatives were and are legal entities that resemble traditional corporations capitalized with money from members who, in this context, happen to be farmers. Cooperatives differ from traditional corporations in some key respects, however. They are organized specifically to benefit the member/owners rather than shareholders, and member/owners participate directly in the ongoing operation and management of the corporation. Unlike stock corporations, profits are shared through patronage dividends, with members receiving a payment based on the proportion of business transacted with the cooperative.

Several different types of cooperatives emerged in the dairy context. A manufacturing cooperative consisted of a group of farmers who banded together to establish a production facility that pooled milk from member dairies into dairy products from which all members would profit. The manufacturing cooperative possessed a number of clear advantages to individual dairy farmers, most obviously the savings gained by operating one collective facility using one set of equipment. In addition, cooperation solved many of the problems inherent in the handling and distribution of milk: since milk spoils rapidly and production is naturally cyclical, a number of farmers working together could breed their cows at different times of the year and even out seasonal production cycles that could prove fatal to an individual farmer's livelihood. Cooperatives also established quality standards for members that improved the overall quality of milk and dairy products. As the farmers of Tillamook, Oregon, already knew, with cooperation, overall consistency of milk and dairy products, and thus their marketability, greatly improved.

Another style of cooperative evolved to facilitate connecting individual farmers or groups of farmers to the marketplace. Oregon's two early cheesemaking cooperatives, the Coos-Curry Cheesemakers Association and the Tillamook County Creamery Association, are both examples of marketing cooperatives in which independent cheese factories, some of them cooperatives themselves, banded together to share marketing. In 1916 Oregon buttermakers followed that model when they formed the Oregon Cooperative Dairy Exchange to market the output of its members and cut out jobbers and middlemen. Perhaps the most significant regional marketing cooperative of this period was formed in 1918 in Washington, when five of the state's manufacturing cooperatives in western Washington—the Skagit County Dairymen's Association, Snohomish County Dairymen's Association, Lewis County Dairymen's Association, Whatcom County Dairymen's Association, and the Enumclaw Cooperative Creamery—formed the United Dairymen's Association of Washington. United enlisted Seattle dairy products broker Umberto Dickey (United's answer to Carl Haberlach in Tillamook) and his Consolidated Dairy Products Company in Seattle to market the group's collective output of products. United eventually settled on the still-familiar Darigold as the collective brand name for butter, cheese, ice cream, fluid milk, and other products produced by its members. Virtually overnight the United Dairymen's Association locked up a large portion of the milk supply in the state, on its way to becoming a Pacific Northwest dairy powerhouse that still exists today.

While many cooperatives were successful, and still-familiar Northwest dairy brands like Tillamook and Darigold are demonstrative of the possibility of ongoing success, the idea of cooperation was not always popular among farmers during the early twentieth century, nor were attempts at organization universally successful. Many dairy farmers felt that by remaining independent they would be better able to take advantage of the highest price offered for milk. "There was some that wouldn't [join]," according to Skagit Valley dairy farmer Emil Youngquist. "Some people [didn't want] to belong to a co-op because they wanted to ship their milk wherever they pleased." The Mt. Angel Cooperative Creamery in Mt. Angel, Oregon, came about in fits and starts after three separate attempts by private entrepreneurs to establish a local creamery had failed. The last proprietor, Mr. Ruidal, left town in the middle of the night, leaving many area farmers without payment for their cream. In 1912, a few area farmers got together and decided to start their own cooperative concern. But the leaders of the fledgling organization encountered considerable community skepticism, as their efforts represented the fourth creamery operation that farmers would be asked to trust, and at that point they were not in a trusting mood. Though the organizers canvassed the countryside by horse and

Mt. Angel Creamery, Mt. Angel, Oregon, circa 1920s. Photo courtesy Mt. Angel Historical Society.

buggy, their efforts proved to be ineffective in signing up members. Perhaps the biggest barrier to joining was financial—few farmers had any money to join and local banks were unwilling to lend startup money to a type of business that had clearly not been very successful in Mt. Angel. As a result the leaders borrowed heavily on their own personal credit to pay members for their cream, pledging their own property as collateral to keep the operation running, but at the same time forging trust in the community for their commitment to the operation's success.

One infamous regional cooperative, the Oregon Dairymen's League, set the idea of cooperatives back forever in the eyes of many dairymen throughout the Pacific Northwest. The league was formed in 1917 among milk producers in the greater Portland area. Shortly after its formation the organization went on the offensive, asserting that milk prices in the Portland area were so low relative to the cost of production that its member dairy farmers were being forced to slaughter cows in order to be able to afford to feed and maintain the rest of their herds. Low milk prices were endangering the city of Portland's milk supply, the society implored, and it was "the customer's patriotic duty to stand by the producer."

Portland area milk distributors challenged the league's tactics and sued the organization in 1918, alleging that the organization's effort to influence milk prices violated the Sherman Antitrust Act. The Dairymen's League retaliated by announcing that it would simply bypass local distributors altogether and set up its own distributing system, and toward that end it purchased a brewery in downtown Portland to serve as a central warehouse and milk plant. This caused considerable internal strife amongst the league's members, many of whom believed this considerable capital investment would bankrupt the organization. The league forged ahead anyway and in the next several years acquired a number of creameries and cheese factories in Oregon as far south as the southern Oregon coast. At its peak the Oregon Dairymen's League boasted a membership of over three thousand farmers and operated twenty cheese factories and five buttermaking plants across the state in addition to its milk plant in the city of Portland.

Soon member farmers began to rebel against their contracts, which mandated that they sell milk solely to the league, and several filed lawsuits alleging fraud and misrepresentation on the part of the organization. Accusations of poor financial management began to surface when it became clear that the organization had expanded too quickly and could not keep up with the massive debt it had accumulated in the process. Dairy farmers in Coos and Curry counties rebelled and began selling their milk to a Nestlé condensed milk plant in Bandon that had opened in 1919. Eventually five hundred and ten Clatsop County dairy farmers resigned *en masse* and formed their own cooperative, the Lower Columbia River Dairymen's Association, citing among other complaints their lack of a voice in the league's affairs. The Oregon Dairymen's League crumbled as membership in other parts of the state, sensing the end was near, also began to break away. By the end of 1921, the league had disbanded entirely.

Kraft Comes to the Pacific Northwest

Today Kraft Foods, Inc., is one of the largest food companies in the world, but it evolved from humbler origins. James L. Kraft, a native of Ontario, Canada, began his career as an investor in the Shefford Cheese Co. in Buffalo, New York. While Kraft was on a business trip to Chicago in 1903, the other partners are said to have abruptly cut him out of the business. Not deterred, Kraft purchased a horse and buggy, bought cheese from Chicago wholesale merchants, and peddled his wares door to door. The business proved successful, and by 1914, Kraft and his brothers purchased a cheese manufacturing facility in Stockton, Illinois.

J. L. Kraft went on to revolutionize the cheese industry with the patent he secured in 1916 for manufacturing a product called "processed cheese." As the name implies, the product was not exactly cheese; the process in processed cheese involved mechanically pulverizing cheddar cheese, applying heat, agitating the product, and packaging the end result in cans . Because the cheese that wholesalers and consumers of the period typically purchased was often of uneven quality and prone to spoilage the consistency and uniformity of processed cheese was very appealing. Processed cheese was very popular with consumers and Kraft went on to manufacture and sell thousands of cans to the retail trade and to the United States government, which in turn shipped them to its troops in Europe during World War I.

With national demand for cheese increasing after World War I, the forward-thinking J. L. Kraft sought to expand his manufacturing base. He saw that the dairying states of the East and Midwest had become saturated with dairy farms and cheese and butter factories and recognized that, to continue to grow, his company would have to look elsewhere. "In order to safeguard our future," Kraft wrote in *Cheesekraft*, the intracompany newsletter, "[we must] look to the West for the extra hundred million pounds of cheese which we ourselves expect to use by the year 1925." Fortuitously, in 1922 Idaho Governor David Davis invited national dairy industry officials including Kraft to tour the state of Idaho and adjacent areas of eastern Oregon. Kraft was greatly impressed by the potential he saw in the developing state. "Idaho has the climate, the high production of feedstuffs per acre and the general soil conditions to make it a leader in dairy products," he told the Boise Rotary Club during the tour. "Idaho has the largest possibilities [for dairying] of any state in the Union." Directly as a result of this trip, several of Idaho's existing cheese factories started shipping cheese to a Kraft processing facility in San Francisco. So convinced was Kraft of the state's potential that he relocated the company's San Francisco operation to Pocatello, Idaho, in 1924, conveniently near the site of the Minidoka Irrigation Project, which had already turned the area into a dairying center in the state. The Pocatello plant was a significant stimulus to the area's dairy economy since it purchased cheese from local factories to produce processed cheese. For a period of time Kraft labels and advertisements announced the Idaho connection proudly, proclaiming: *Kraft Cheese—New York, San Francisco, Pocatello*. Idaho's cheese production increased 330 percent between 1920 and 1925, reflecting the stimulus provided by Kraft's entry into the state.

Kraft was innovative in its methods of expanding into the Pacific Northwest. The Laabs family, already well established in Wisconsin cheesemaking circles, announced their intent to expand into Idaho in 1922, an effort financed by Kraft. J. W. Laabs promised to open facilities at Jerome and Burley as well as re-opening closed plants in Rupert and Paul in the Minidoka area, adding that if Kraft didn't follow through on the financing there were other Midwestern cheese operations that were ready to do so. The Wisconsin company went on to operate seventeen cheese plants in Idaho and the entire output from every one was sent to Kraft's Pocatello processing facility. Laabs sold its Idaho operations back to Kraft in 1927. Utah-based Nelson-Ricks and Mutual Creamery Co. also expanded their operations into Idaho during this

Red Rock Cheese Co.

Like many early twentieth-century housewives, Helen West of Tigard, Oregon (south of Portland), often made cottage cheese at home; recipes for cottage cheese appeared everywhere in ladies' magazines, cookbooks, and newspapers of the period. But Mrs. West's cottage cheese was different, at least in part because of the entrepreneurial spirit of her son Harry, who had the idea of selling it on his milk route for extra money. Harry West found that his mother's cottage cheese was a hit with local residents and Helen West began making increasingly larger batches. As her cottage cheese grew in popularity, her husband, Charles, a former trolley driver in the city of Portland, worked out a unique distribution deal: he convinced the crews of the Oregon Electric trolley line, which ran along the back of the West property, to carry his wife's cheese to customers. West built a small platform on his property adjacent to the tracks, and at an appointed time the trolley would slow just enough to allow him to slip his wife's cheese on board. Such was the demand that the family eventually constructed a separate building behind their house devoted exclusively to making it.

Red Rock Cheese Company was formally established in 1919 and the company built a factory in Tigard. Harry West became president and led Red Rock on a course of rapid expansion during the 1920s that included the acquisition of two additional plants in Washington and in California. The Wests continued to operate Red Rock until 1929, when the fortunate family sold the company to the expansion-minded Kraft-Phenix Company just months before the stock market crash. Kraft operated the Tigard factory for several decades before finally closing the plant in the 1950s.

time; by the mid-1920s Nelson-Ricks operated eleven cheese plants in the state and Mutual operated four.

Clearly Kraft was focused on growth. In 1928, Kraft acquired the Phenix Cheese Company, a large cheese manufacturer located in Pennsylvania and creator of the well-known brand Philadelphia Cream Cheese. The merger came about in part because of a simmering dispute between the two companies over various patents Kraft had claimed for processed cheese. After the merger the buying spree began in earnest: Kraft purchased dairies and cheese factories across the nation including Red Rock Cheese Co. in Tigard, Oregon, a large

Nelson-Ricks Cheese Factory #5, St. Anthony, Idaho, circa 1930. Photo courtesy Special Collections and Archives, University of Idaho, Moscow, Idaho.

producer of cottage cheese that also owned plants in Kent, Washington, and Tulare, California. By 1930, Kraft itself had been acquired by National Dairy Co., making it one of the largest companies, dairy or otherwise, in the United States. The company grew large enough that its economic influence extended beyond dairy products; during the 1930s lumber mills at Oregon City, Oregon, and Cathlamet, Washington, produced wooden packaging for Kraft products and Kraft opened a large distribution warehouse in Seattle in 1936.

Goat's Milk Cheese Arrives

The increasing availability of goat's milk during the early decades of the twentieth century led to the introduction of domestically produced goat's milk cheeses. As early as 1914 at least two farms in Utah were producing and selling goat's milk cheeses, one in Lark and one in Bingham. In the Ettersburg Valley of Humboldt County in northern California, farmers organized the North Counties Milk Goat Association in 1922. Dairy goat farmers of the Pacific Northwest also started down the path of making cheese. One of the organizers of the Oregon Dairy Goat Association, Dr. J. Murphy, was a stockholder of a new enterprise called Portland Goat Dairy, formed in 1920. Portland Goat Dairy sold goat's milk in the Portland area for several years and expressed its long-term goal of making cheese, though these plans never seem to have been realized. In 1921 in Yakima, Washington, Gustos Kontos reportedly took initial steps toward developing a goat cheese industry there like that of his native Greece, but his plans were stymied when the Greek government would not allow him to export Maltese goats to the United States.

Winnie and Jay Branson started one of the Pacific Northwest's earliest goat cheese operations in the Willamette Valley west of Salem, Oregon, in 1925. The Bransons kept a mixed herd of around a hundred and twenty-five Toggenburg, Nubian, Alpine, and Saanen goats, all of which they knew by name. They combined cow's milk with their goat's milk to produce a Roquefort-style blue cheese that they sold to hotels and stores in the Portland area and as far away as New York. Not only were the Bransons among the first to produce goat's milk cheeses in the Pacific Northwest, they were also among the first commercial blue cheese makers in the United States.

The Branson farm was tucked in the foothills of Oregon's Coast Range near Falls City, an area with a number of natural springs around which the Bransons constructed their cheesemaking operation. They stored milk in cans that they submerged in the cool springs until they were ready to make cheese. The salting room was built so that water flowed over the floor at all times, maintaining a constant temperature of about 55 degrees. Spring water was piped over the top and down the sides of the ripening room as well as over the floor in order to mimic conditions in a damp cave like those of Roquefort, France. A visiting reporter described the ripening room as having "a cheesy odor not quite pleasant at first," though the cheese that came out of the room was "of goodly flavor." The Bransons aged their cheeses about seventy days, then wrapped each cheese in wax paper lined with aluminum foil, and packed them in a box filled with sawdust for shipping.

The Bransons seem to have had ambitious plans to expand their reach into national markets. In 1926 they had enlarged their facilities and began processing milk purchased from five hundred area goats to make their cheese. The expansion effort did not prove fruitful, however, and accounts from a few years later indicate that they had downsized back to a smaller farmstead operation. By the 1930s the Bransons were out of the cheese business entirely; the onset of the Depression may have doomed their ambitious cottage industry.

M. P. Eggers, one of Washington's dairy goat industry pioneers, finally got into the cheese business himself, evolving his Encaria goat's milk operation into Briar Hills Dairy in North Bend, Washington, outside Seattle, in 1928. Eggers was a dogged marketer who focused heavily on the health angle; he touted his Viking Cheese in early ads as "an organized food that gave the sturdy ancients from the North STAMINA ENDURANCE ENERGY" (*emphasis in the original*). One of Eggers' best-selling products was a powdered goat whey product called Whex. Whex became very popular, especially after prominent California health guru Dr. Bernard Jensen began promoting it to his patients and followers during the 1950s. The majority of Eggers' business was done via mail order through ads he placed in small-circulation magazines like *Dairy*

Goat Journal and *Organic Gardening* and a variety of regional goat-focused publications from across the country.

Another regional goat cheese factory was founded in East Stanwood, Washington, in Snohomish County, north of Seattle, in 1928. The Western Goat Products Company, which the *Seattle Daily Times* claimed was "America's Only Goat Cheese Factory," made eight hundred pounds of cheese daily during its first year of production. Western Goat Products hired cheesemaker John Castrilli, whose father, Louis Castrilli, operated a cheese plant to the east in nearby Hamilton, and focused on producing Norwegian style gjetost and a cow's milk cheese it called bamm ost, sold under the Maid of Norway brand. According to the *Stanwood Times*, the outfit maintained its own herd of Saanen goats at a ranch on adjacent Camano Island in Puget Sound.

The local *Stanwood Times* was quick to report the company's progress throughout 1928 ("ranks among the best" was one presumptive early cheer), but between the lines, its stories suggest that President A. G. Clark and George Munson, secretary/treasurer, had their hands full with the new enterprise. Securing a consistent supply of goat's milk seems to have been one of the biggest challenges; Clark spent a fair amount of his time canvassing the local community and addressing members of local chambers of commerce, encouraging them to develop dairy goat farms. The company later branched out into additional styles of cow's milk cheeses like edam and brick cheese, perhaps an indication of the difficulties the company encountered in securing enough goat's milk to maintain the operation. The move may have also been indicative of a weak market for goat's milk cheese; though the Skagit Valley was home to many settlers of Scandinavian origin to whom Norwegian-style cheeses might appeal, the products were probably unfamiliar to those in larger markets outside of the area.

In 1929 George Munson purchased Western Goat Products outright and installed a new cheesemaker, Richard Mus, a native of Holland. Monson changed the name of the company to Dairyland Cheese Company the same year, shifting the branding focus of the company away from goats exclusively. The shift toward cow's milk was risky, however, because Stanwood was also home to a large Carnation condensed milk plant that had a well-established relationship with area dairy farmers. Though Munson's factory was said to have paid farmers a higher price for their cow's milk than the condensary, that practice would have been sustainable only with consistent and increasing sales.

Industry Innovations

During the early decades of factory cheese production in the United States, most of what was produced was cheddar cheese. In the cheese industry cheddar is, strictly speaking, a process of making cheese rather than a flavor profile. Cheddar production became widespread because the process of making it proved easily adaptable to a mass production factory environment; the cheddar process produces a consistent, uniform product that ages quickly and remains generally stable during shipping and handling compared to other types of cheese. But even as overall United States cheese production increased steadily during the late nineteenth and early twentieth century, the United States continued to import many millions of tons of other styles of European cheese. The high volume of imports was in part because exact mechanisms by which many European-style cheeses were produced eluded cheesemakers and scientists, and for many years familiar styles of imported cheese such as the blue-veined cheeses of France, Italy, and Denmark simply could not be replicated in commercial quantities in the United States.

That state of affairs began to change in 1895, when the federal government created the Dairy Division of the Bureau of Animal Industry. The founding of a division specifically devoted to the study of dairying in all its aspects was a significant sign of the growing importance of the industry and its place in the nation's economy. The Dairy Division established a number of research facilities on the East Coast, including a working dairy farm in Beltsville, Maryland, with two hundred and fifty head of Jersey and Holstein cattle, a leased cheese factory in Grove City, Pennsylvania, and a laboratory at Storrs, Connecticut, where, among other things, the government maintained a herd of dairy goats. Scientists conducted a wide range of studies in animal husbandry, milk safety, and the manufacturing and processing of milk and dairy products. In addition to overseeing the eradication of tuberculosis in the U.S. cattle population, the bureau conducted research in the rapidly growing field of microbiology and the science of milk, cheesemaking, and cheese ripening. Their efforts contributed directly to the growth and development of the cheese industry in the United States, and in particular, to the understanding of specialty cheese mechanisms and processes that cheesemakers in the United States eventually began to exploit.

Blue Cheese in the Pacific Northwest

In the paper *Fungi in Cheese Ripening: Camembert and Roquefort*, government microbiologist Charles Thom began to crack the heretofore mysterious source of the "blue" in blue cheese. Thom isolated the mold from a specimen of Roquefort cheese, noting that the microorganism was the sole agent responsible for the flavor and texture of Roquefort cheese. "There seems to be no doubt," he wrote, "that it will be possible to develop methods of making and ripening that will produce the Roquefort type of cheese in the United States." By 1919, government researchers at the Grove City, Pennsylvania, Experiment Station had constructed a facility that recreated temperature and humidity conditions present in the caves at Roquefort and were producing a reasonably good cow's milk blue cheese in some quantity. The Bransons, makers of mixed-milk blue cheeses in Falls City, Oregon, in the 1920s, ordered their supplies from the Grove City Experiment Station.

Across the nation scientists began to apply themselves to conquering the mysteries of blue cheese production. Researchers at the University of Minnesota and at Clemson Agricultural College in South Carolina experimented with the aging environment in area caves (in Minnesota, caves along the Mississippi River created artificially by the extraction of silica; in South Carolina, an unfinished railroad tunnel) that they hoped would replicate the natural cave conditions in Roquefort, France. In 1936, Iowa State University researchers perfected and patented a process for making blue cheese with homogenized cow's milk that they licensed to the local Maytag family, who went on to commercial success producing a product they called Maytag Blue. Through the combined efforts of government and university scientists, by the late 1930s large-scale commercial blue cheese production began in earnest. In 1949, 8.1 million pounds of blue cheese were produced in the United States.

The advent of domestic production of blue cheese in the United States was nurtured by a number of other factors as well. Import shortages during World War I along with the turbulent European politics of the 1930s created a market gap for European-style cheeses in the United States. Blue cheese also held the potential to be sold for a premium price as compared to commodity cheddar, a fact that must have looked very attractive to a smaller producer with tight margins. The product also surely appeared to hold promise as a potential product line; the novelty of a domestically produced blue alone might attract curious buyers who would (hopefully) continue to buy the product for its quality as well as for its lower price.

Langlois Blue Star Cheese Co. (Langlois, Oregon)

Curry County, Oregon, in the far southwestern part of the state, adjacent to the California border, had long been a center of dairy farming. In fact, so many dairy farms sprang up in the area around the present-day town of Langlois that locals once referred to the town as 'Dairyville.' The one-thousand-plus-acre Star Ranch was well known in the region around the turn of the twentieth century as a productive dairy operation run by E. W. Catterlin, formerly of Tillamook. As early as 1911, Catterlin milked a hundred and fifty dairy cows, a huge herd during a period when ten to twenty cattle was more the norm and hand milking was still standard practice. Catterlin developed a cheese factory

Workers package blue cheese at the Langlois Cheese Factory, Langlois, Oregon, 1955. Photo courtesy Oregon State Archives.

on the ranch and shipped his cheeses by rail to Portland. During the first two decades of the twentieth century a number of cheese factories came and went in the area, including the Ocean View Dairy and the Cloverdale Cheese Factory (not to be confused with the Cloverdale factory in Tillamook County). In 1914 Langlois dairy farmers organized their own cooperative, the Langlois Dairy Products Company, which for several years operated a cheese factory that produced upwards of seven hundred pounds of cheese a day. In 1919, Milt "Bullhide" Moore built two cheese factories, one at Ophir, north of Gold Beach, and another about twenty miles to the south at Pistol River. Moore's factories sold butter and cheese to Chicago-based Swift and Co. wholesalers, which distributed the products under the Brookfield label.

Danish immigrant Hans Hansen arrived in the south coast area in 1915; his brother, Schmidt, was owner of a sporting goods store and shooting gallery in Bandon. Hansen rented the Star Ranch in 1916 and by 1926 had purchased it outright. Hansen operated a successful cheese factory there for several decades and became a member of the Coos-Curry Cheesemakers Association when it formed in 1916. In 1940, Hansen began experimenting with blue cheese working in conjunction with fellow Danes Verner Nielsen of Iowa State College (who had participated in the development of Iowa State's patented blue cheese process) and experienced local cheesemaker Leon Warming. The men perfected what came to be called Langlois Blue Star Cheese that same year. Langlois Blue was available for sale in Los Angeles by mid-1941; one reviewer there called it a "superbly perfect cheese," and rated it better than other domestic blues then being produced in Minnesota, Illinois and Wisconsin. A reporter who visited the factory described Langlois blue this way:

> Langlois cheese has delicious piquant flavor and leaves a fine mild after-taste as does French Roquefort, but is not as dry and crumbly as that cheese. The texture is similar to English Stilton: smooth, easy to cut. The color is neither white nor yellow, but an appetizing ivory.

By the 1950s Langlois Blue was available for purchase in New York City and received several favorable mentions in the *New York Times*. The Blue Star Cheese Company developed into a popular tourist attraction for drivers along the newly constructed Highway 101 on the Oregon coast. Hansen also sold a good portion of cheese by mail order and through southern Oregon gift retailer Harry & David—to the tune of over half a million pounds of blue cheese worldwide by 1956.

Tiny Langlois' cheesemaking fortunes abruptly ended when Hansen's cheese factory burned down in a boiler fire in April of 1957, an event significant enough to warrant mention in Portland and Seattle newspapers. Though Hansen told reporters that the loss was fully insured and he intended to rebuild the factory, a new facility never materialized. An attempt was made to revive the legacy of blue cheese production in Langlois in 1968, when a company calling itself Blue Cheese Processors of Langlois, Inc., issued a stock offering directed at Oregon residents who were urged to buy into the newly formed company at $2 per share. A rusted sign remains at the factory's former location just off Highway 101 in Langlois, a melancholy memorial to this once nationally known blue cheese company.

Guler Cheese Co. and Black Rock Cheese Co. (Trout Lake, Washington)

Klickitat County's earliest settlers, many of Swiss and Scandinavian origin, brought their dairy and cheesemaking experience with them to the region. Swiss native Joseph Aerni drove cattle from Portland across the Columbia River to Washington and northeast into the Trout Lake Valley, just east of Mt.

Adams, in 1885 and started a dairy farm. Once established, Aerni set up a section of his woodshed as a cheese production facility. His swiss-style cheese became so popular that he was eventually able to save enough money to build a formal cheese house equipped with an eighty-gallon iron kettle that he hung over an open fire.

Natural caves are plentiful in this mountainous volcanic region, and the area's early settlers took advantage of them to store butter and meat. During the 1920s prominent area businessman and landowner Wade Dean had the idea to develop a business storing potatoes in a large cave on his property, though the relative remoteness of the cave as well as the practical difficulty in getting potatoes in and out of the cave ultimately doomed this venture. Several years later, Dean's daughter Dalia married Homer Spencer. As the story goes, Spencer became interested in the possibilities of the Trout Lake caves for making blue cheese while convalescing from a broken leg. The location must have seemed like a sure thing, in a mountainous region like the Roquefort caves in France and at nearly the same altitude and latitude. Working in consultation with Dr. L. A. Rogers of the USDA, Spencer nurtured a wheel of cheese sent from the government's Grove City Creamery research facility in Pennsylvania in the caves on the Dean family property. After months of aging, the cheese was, according to Spencer's description, "as mellow as butter, daintily marbled with blue-grey mold, exactly correct as to texture." Wade Dean leased the cave property to his son-in-law for a period of thirty years, and by the early 1940s the Guler Cheese Co. was up and running.

The Guler Cheese factory consisted of a barn outfitted as a cheese facility with adjacent buildings serving as salting and packaging rooms. The caves, located at the bottom of a steep set of stairs that extended fifty feet underground, were outfitted with wooden racks on which the cheeses were aged for eight months. During that period the cheeses required a considerable amount of

Guler Cheese: "It's the Cave." Label courtesy John Shuman

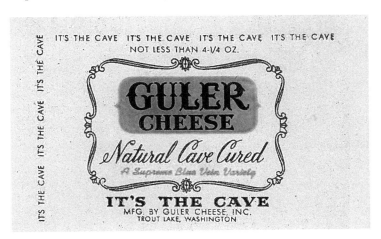

maintenance including periodic turning and needling to allow air to penetrate the surface, encouraging mold growth. The cheeses were wrapped in foil during the final months of aging. Spencer worked hard to sell the romance of his cheese's unique aging environment—the Guler Cheese label declared in bold lettering, "It's the Cave." Surely he was aware of the competition coming out of Langlois in southern Oregon, a cheese that had already made great strides in national popularity—but was not aged in a natural cave.

At this point the tale of Guler Cheese takes on a bit of a melodramatic flourish. Spencer's cheese business took a turn for the worse after a messy divorce in which both Spencer and his wife accused each other of cruelty; we know some of the prurient details because Dalia Spencer appealed the division of property all the way to the Washington Supreme Court. Because Washington is a community property state, the marital property up for division included the cheese operation and thirty-year cave lease. At trial, the judge awarded Homer Spencer the cave lease and equipment, but the Washington Supreme Court determined in a 1946 ruling that Dalia Spencer was entitled to a one-half share of the cheese business.

After the divorce the Dean family withdrew their financial backing from the company, though Spencer apparently continued to have access to the cave. He scrambled to find new financing, securing among other things a community development project grant sponsored by the University of Washington. He also renamed his company Black Rock Cheese Company. Spencer seems to have had hopes of distinguishing this new brand by making the Black Rock product entirely, or nearly entirely, of goat's milk—a cheese the *Oregonian* newspaper proclaimed in a headline "A New Cheese" that was nearly an exact duplicate of French Roquefort. Regardless of the new recipe, the name change surely created an enormous marketing challenge and the business slid into a decline from which it never recovered. In 1956 Spencer's brother Walter and other investors took over the faltering operation in an apparent attempt to revive it, reclaiming the name "Guler Cheese," but they went out of business shortly thereafter.

Family patriarch Wade Dean eventually built an A-frame home over the entrance to the former cheese cave (the structure remains there today) and replaced the wooden staircase that leads down into the cave with a more durable metal one. He was said to have plans to turn the cheese cave into a gallery of sorts with an elaborate display of busts of friends made during his long career, but he passed away before the project was carried out. Intrepid spelunkers can still explore the abandoned cheese cave and its deteriorating wooden racks, all that remain of Homer Spencer's ambitious blue cheese empire.

Darigold Blue Cheese (Menlo, Washington)

Menlo is a small town in Pacific County, Washington, in the far southwestern part of the state near the Pacific coast. Like many areas of the Pacific Northwest, the area's economy thrived with the surging logging industry for many decades and at one time the hum of sawmills was as common as the sound of breaking ocean waves.

Near Menlo in the town of Lebam, dairy farmer C. P. Dobler, a native of Switzerland, operated a dairy farm and made cheese under the Helvetia brand. Dobler and several other area dairy farmers started a cheese factory in Menlo in 1900 but that attempt fizzled quickly. Dobler continued dairy farming and daughter Estella took over as cheesemaker in later years, producing what was known as Mayfield Swiss Cheese on the family farm; Estella was also known as an exceptional artist and was said to have decorated her cheese room with her art.

In 1914 another group of businessmen formed the Willapa Valley Cheese Company and built a new plant in Menlo to house the operation. That factory burned down in 1917 (a common fate of many cheese factories during the period as they were typically constructed of wood) but was rebuilt a few years later. In 1921, the Willapa Valley Cheese Company joined the Lewis County Dairymen's Association based in Chehalis, about forty-five miles to the east. Over time, the Lewis County Dairymen's Association moved to consolidate area cheese production to its large Chehalis factory. In 1947 the cooperative re-tooled the Menlo plant at considerable expense to enable it to produce blue cheese, which was sold under the Darigold brand. Raymond Falk, an expert from Wisconsin, was recruited to supervise the operation. The idea of making blue cheese as a last-gasp survival strategy proved short-lived, however, and the retooled Menlo plant was open for just five years before closing for good in 1952.

Rogue River Valley Creamery (Central Point, Oregon)

The southern Oregon interior had been populated by dairy farms as soon as early as settlers began to stream there, following the Applegate Branch of the Oregon Trail to points south of the Willamette Valley. As in other parts of the state, a number of creameries and cheese factories emerged along with the growing population, among them Fort Klamath Creamery, which opened in the late 1890s, and the Rogue Valley Creamery, which first began operations in Grants Pass in 1913. By 1930 there were two cheese factories in the Klamath Falls area and A. Woodrich was poised to open the Ladino Cheese Factory in Eagle Point north of Medford.

Sonoma Valley Cheese Factory, Central Point, Oregon, plant, circa 1938. Photo courtesy Rogue Creamery.

In 1932 a group of investors approached the city of Central Point about starting a cheese factory. The city council agreed to give the group the site of an old brickyard near the railroad tracks for free in exchange for their investment in the city's economy. Rogue River Cheese & Produce Company opened with great fanfare in February of 1933, managed by William Churchill, with H. H. Biberstein as lead cheesemaker. Although the local media reviewed the company's cheese favorably, the organizers encountered significant skepticism from area dairy farmers and had trouble obtaining milk. By September, the factory had closed and the owners declared bankruptcy. Ed Seufert of The Dalles, the largest creditor, resurrected the operation in an attempt to recover his investment. The factory seems to have limped along until Celso Viviani and Gaetano "Tom" Vella, owners of the Sonoma Valley Cheese Factory in Sonoma, California, purchased the Central Point operation from Seufert in 1935 with financing provided by J. L. Kraft. Viviani and Vella operated the plant as the Sonoma Valley Cheese Factory, Central Point Plant, until the mid 1940s, when they dissolved their business partnership; Tom Vella retained ownership of the Central Point cheese plant and renamed it Rogue River Valley Creamery. Like many smaller cheese plants of the era, the plant manufactured cheese under contract for Kraft throughout World War II, producing five million pounds of cheddar a year that went directly to the war effort. Because of the dearth of available labor due to the wartime draft, Vella hired women workers to keep the plant running.

According to Ignazio "Ig" Vella, Tom's son, during the 1950s representatives of the Borden Company, for whom the factory was producing its cheese under contract at the time, suggested that the plant look into producing blue cheese, since the price of the blue cheese coming out of Langlois was becoming so high. Tom Vella and his wife took a trip to Roquefort, France, in 1956 to see how it was done. As luck would have it, the manager of the Roquefort caves was Italian and more than happy to show fellow countrymen the ins and outs of its production. The Vellas toured dairy farms, cheese factories, and the limestone caves of Mont Combalou; Vella returned to the United States and constructed an aging facility at Central Point designed to mimic cave conditions in Roquefort, France. By 1957 (coincidentally, or perhaps fortuitously, the year the Langlois blue cheese factory burned down) Rogue River Valley Creamery began producing its first blue cheeses. Rogue Creamery is the sole survivor of the Pacific Northwest's early blue cheese producers, and continues to make several styles of blue cheese today. According to the creamery, Oregon Blue is still made using Tom Vella's original recipe.

Canned Cheese Boom and Bust

Prior to the advent of refrigeration, cheese manufacturers and retailers faced a significant challenge in storing and transporting cheese. While aged cheese typically develops a relatively impervious rind, variations in manufacture, handling, and storage led more often than not to spoiled, broken, or moldy cheeses that were of poor quality and could not be sold by retailers. Some manufacturers packaged cheese in foil-packed wooden boxes that solved some of these problems. Canning was viewed at the time as another potential solution; by placing a perishable product in a tin can and vacuum sealing it a manufacturer could preserve the item indefinitely. It was only a matter of time before canning technology found its way to the cheese industry.

In the dairy world, J. L. Kraft had already found considerable success manufacturing processed cheese and packaging it in tins. But packing regular, unprocessed cheese in cans proved to be more of a challenge. As early as 1904, Emil Pernot at Oregon Agricultural College studied the possibility. While Pernot had some success, his results were uneven and he found that cans of cheddar cheese were prone to swelling and bursting, especially at high temperatures. The technology was promising enough, however, that government and university scientists around the country continued to work on perfecting the process. During the 1930s, USDA scientists developed a can with a built-in valve that allowed off-gassing as the cheese aged in the can; the resulting product was stable enough to be produced and marketed commercially. In 1933, the government's Grove City, Pennsylvania, cheesemaking facility released ten thousand cans of cheese in an effort to test the market for the product. The prospect of a viable canning process for cheese was promising enough that the *National Butter & Cheese Journal* declared it "one of the most revolutionary developments that has ever taken place in the [cheese] industry."

A number of large commercial manufacturers seized on the new technology, including the Atlantic and Pacific Tea Co. (A&P) and Land o' Lakes, and canned cheese began to appear in retail stores around the country. According to contemporary accounts, Portland, Oregon, was the first test market in the nation where canned cheese was specifically branded and promoted as a unique product. Several manufacturers under the United Dairymen/Dari-gold cooperative umbrella, including Interstate Associated Creameries, sold canned cheese in the Portland area in 1935. Interstate went all out, conducting a naming contest (Miss Ellen Dawson took first with "Golden Harvest") and awarding prizes for recipes using the product (Savory Sandwich Buns won the top cash prize of $25). In 1938 the company shipped several cases of Golden Harvest to Philadelphia in an effort to test the market for its product on the eastern seaboard.

Meanwhile, at Washington State College in Pullman, Dr. Norman Golding was working independently, in conjunction with officials from the American Can Company, to develop a viable canned cheese product. By 1937, Golding had perfected a cheddar cheese recipe that could be successfully packaged in a regular (non-valved) can. What made Golding's cheese different was that he used an adjunct bacterial culture that, in combination with the cheddar culture normally used in the cheesemaking process, limited the production of carbon dioxide, which had contributed to the exploding can problem. Golding's cheese was eventually dubbed "Cougar Gold" (the Washington State University mascot is a cougar, and the 'gold' is for Golding). The product continues to be manufactured and sold today in Golding's signature cans, using milk from the school's dairy herd managed by Animal Science Department students. Dr. Golding went on to develop and patent a technique for canning blue cheese as well.

While canned cheese was *the* cutting edge packaging technology of the 1930s, in the end it turned out to be a passing industry fad. Some processors balked at taking up the technology due to the expense of outfitting their factories and others objected because the canning technology was not proprietary and could be used by any manufacturer. More significantly, canning lost considerable traction in the 1940s as Pliofilm and successor plastics took center stage, products that allowed manufacturers to age and package cheese much more simply and efficiently than in heavy, bulky metal cans. Canned cheese is still a viable technology, however, and continues to be produced in and for areas of the world where refrigeration and transportation are less reliable.

Another War, More Cheese

As during World War I, the advent of World War II was a great stimulus to the Pacific Northwest economy. In Portland and just across the Columbia River in Vancouver, Washington, Kaiser Shipyards kept workers busy producing ships and aircraft carriers; in Seattle, Boeing once again ramped up production, manufacturing thousands of aircraft for the war effort, made of metal this time around instead of spruce. Authorized by the Lend-Lease Act of 1941, the U.S. government began shipping food, equipment, and weapons to European countries fighting the Axis powers. Dairy farmers benefited from government purchases of staggering amounts of dairy products. Kraft, Borden, and other national industrial-scale manufacturers led the way in selling dairy products to the government, using products from their own plants as well as leasing production facilities and purchasing cheese directly from smaller producers across the country. By mid-1942 the United States government had purchased over three hundred and sixty million pounds of cheese from the nation's cheese factories to send to Europe. Total U.S. cheese production increased 21 percent between 1940 and 1941 and another 16 percent in 1942.

According to manager Carl Haberlach, Tillamook sold a total of 1.4 million pounds to the United States Government in 1941. Oregon's overall cheese production increased more than 35 percent during the first three months of 1942, an increase attributable at least in part to the fact that creameries not previously outfitted for cheese production had done so in anticipation of wartime need. The state's cheese factories went on to set a cheese production record, making over sixteen million pounds that same year. Idaho cheese production reached over nineteen million pounds in 1942.

During this period the federal government established mandatory set-asides for cheese producers; rates varied throughout the war period but as much as 70 percent of cheese production was required to be held back from commercial sale by producers for government purchase. Regulations also restricted the production of so-called "foreign-type" cheeses such as brie in order encourage the production of cheddar. An elaborate labyrinth of programs evolved to support the massive expansion of the dairy industry, including the diversion of corn and other grains for use as cattle feed on ranches and dairy farms and farm labor assistance via the U. S. Crop Corps and its Women's Land Army and Victory Farm Volunteer units.

The Lend-Lease program siphoned off a significant amount of the nation's food production, creating inevitable shortages. In contrast to the situation on the domestic front during World War I, strict food rationing was a fact of life in the United States during World War II. Cheese became a precious commodity and was restricted along with other staples like meat, flour, sugar, and cooking oil. Formal cheese rationing began in the spring of 1943. American consumers were not restricted to a specified amount of cheese, but rather were allocated a certain number of points per household (the number of points varied over the rationing period) and could use the points on rationed items. At first cheese rationing applied only to aged cheeses, but was extended later in the year to include such items as camembert, brie, and blue cheeses.

Because meat was more heavily restricted than milk or dairy products, cheese—especially cottage cheese—became the protein of choice for thrifty housewives. Newspapers dispensed advice about thrifty meal-planning options using point allocations as effectively as possible. Americans resigned themselves to cheese and pimento sandwiches, cottage cheese salads, and fancy-sounding rarebits made of melted cheese for the duration of the war. Home cheesemaking even began to make a comeback as newspapers and magazines printed recipes for easy-to-make soft cheeses, highlighting the activity as a patriotic act.

As the war drew to a close, stores of rationed products, no longer being shipped abroad, started to accumulate in local warehouses. By mid-1945, Portland wholesale merchants began lobbying for a ration holiday. "Not only is cheese backing up at the warehouse," asserted one retailer, "but some of it is spoiling on grocery shelves." Seattle wholesaler Frye & Co. joined the chorus, warning that two hundred thousand pounds of cheese in its warehouse were in serious danger of spoiling. By early September of 1945, the Office of Price Administration declared cheese to have a point value of zero, effectively eliminating it from the ration list. Though the Allies declared victory in September of 1945, food rationing continued in the United States well into 1946.

Goat cheese production gained some momentum during World War II; since neither goat's milk nor goat's milk cheese fell under the auspices of the Office of Price Administration, production and consumption were essentially unregulated, creating a window of opportunity for the goat dairy industry. Goat farmers wasted no time situating themselves as an ally to the war effort; The *Dairy Goat Journal* proclaimed "Goat Milk: A Weapon For Victory," on its front cover in April of 1943. In a statement printed in the same issue of the *Journal*, the USDA addressed dairy goat farmers directly, noting pointedly, "Milk production by goats, normally a local food resource, assumes national significance during wartime, with the attendant need for food production on a vast scale. In total war every pound of food counts." The *Dairy Goat Journal* got into the act itself, initiating what it called a "Victory Buck" program that encouraged goat breeders to donate male goats of good lineage to those lacking the means to breed their female goats. Several of the nation's larger goat dairies gained a widespread reputation for their cheese during this period, including Itasca Goat Dairy of Dupage, Colorado, which produced a romano-style cheese and Ozark Capri Cheese Co. of Harrison, Arkansas, whose eponymous Ozark Capri cheese appeared on New York restaurant menus.

The Cheese Factory at Tenas Illihee

Willard N. Jones was a civil engineer by training who operated a car dealership, served as a representative in the Oregon State Legislature, participated in the lucrative timber trade, and owned part interest in a farm and orchard with his brother. Jones is perhaps best known, however, for a series of early twentieth-century land speculation scandals that drew national attention. In one, Jones and several associates were accused of paying Civil War Veterans to homestead on the Siletz Indian Reservation on the Oregon coast. Jones was convicted in a sensational Portland trial in 1905-06 but later received a pardon from President William Howard Taft after allegations of jury tampering surfaced.

Jones went on to operate a dredging company with several partners with plans to develop and sell reclaimed tidelands along the Lower Columbia River near Astoria, Oregon. Toward that end, the company purchased Tenas Illihee Island, one of several large islands near the mouth of the river noted by Lewis and Clark as they made their way toward the Pacific over one hundred years earlier. In its original state, the island was said to have been overgrown with brush, trees and weeds and was covered almost entirely with water at high tide. Willard Jones purchased the reclaimed island from his business partners and established a large farm, dairy and family retreat on the island. Resident workers managed the farm and milked the large herd of as many as 250 cows. Milk was transported in ten-gallon cans by boat across the Columbia to Clifton, Oregon every morning, then by rail to the local creamery in Astoria.

In 1920, Jones' son Robert took over day-to-day management of the island farm and started a cheese factory that operated through the 1930s. Tenas Illihee brand cheese was distributed locally. Robert Jones eventually assumed ownership of the island from his father and during the 1940s he phased out the dairy and cheesemaking operation in favor of raising beef cattle, a pragmatic choice during World War II when meat brought high prices. Over the years, however, management of the flood-prone island farm became increasingly onerous. Robert Jones sold the island in 1961 and the property eventually became a wildlife refuge, now known as Tenasillahe Island. Only scant traces remain of the farm that once thrived there.

Cows coming in from the fields at milking time at Jones Farm on Tenas Illihee island, near Brownsmead, Oregon. Photo courtesy Jessie Jones.

A number of goat's milk cheesemaking operations sprang to life in the Pacific Northwest during the war. Goat farmer O. V. Breese of Pacific Cheese Company in Estacada, Oregon, was producing a cheese described as a "hard type or grating cheese used mostly by the Italian trade" in 1944 and by his own account expected to produce blue cheese in the future. The Oregon Goat Dairy Breeders Association of Portland formed a dairy goat cooperative in 1944 that sold milk and aspired to eventually produce cheese as well. Meanwhile, on the southern Oregon coast near Brookings, the Chetco Cheese Association made a goat's milk cheese known as Chetco Cheese.

Perhaps the most ambitious goat cheese factory to open in the region during this period was the Washington Mountain Cheese Factory in Stevenson, Washington, about forty-five miles east of Portland on the Washington side of the Columbia River. Picturesque Skamania County in the Columbia River Gorge had become something of a center of dairy goat keeping and breeding and numerous small goat farms dotted the region from Camas to White Salmon. Around a dozen area farms supplied the milk of about a thousand animals to the Stevenson factory, which opened in 1940. By 1946 the factory was handling three to four hundred gallons of milk daily, just under half of its total capacity. Its primary products were feta and a cheese called migita that plant manager Earl Harrah described as a cheese of Italian origin made from the whey by-product from the plant's feta production. According to Harrah, all of the factory's output was sold for use as what he called "cooking cheese" in the Midwest.

The period spanning World War I and World War II marked a period of continued expansion of the cheese industry in the Pacific Northwest. Cheese production grew increasingly larger scale as the region's cheesemakers began to find ready markets for their products up and down the west coast. While the war efforts were a big boon to producers of dairy products, the inevitable postwar production rampdown forced many small operations out of business. By the 1940s dairy farmers were either sending their milk to regional factories or had banded together to form local cooperatives. The cooperatives, in turn, grew progressively larger in order to compete with large national brands like Kraft and Borden that raked in large profits during both wars. As the larger producers grew even larger, they established networks of factories and distribution infrastructure across the nation that further entrenched their powerful operations. Cheese had evolved into a highly profitable, mass-produced commodity.

The Tillamook County Creamery Association opened a brand new, state of the art cheese factory in 1949. Photo courtesy of Tacoma Public Library.

Chapter 5
The Mass Production Era

In November of 1949 a newly constructed cheese plant opened in Tillamook, Oregon. Touted in the local press as the "largest in the Pacific Northwest," the modern facility was built over a period of two years and cost $1.5 million. The sprawling factory included a storage warehouse capable of holding three million pounds of aging cheese, equipment for manufacturing a variety of dairy products including cheddar cheese, cottage cheese, ice cream, and butter, as well as a milk bottling plant. The factory even boasted its own independent water supply drawn from a reservoir created by damming nearby Cole Creek. Retired manager Carl Haberlach, so instrumental in the cooperative's early growth and success, died suddenly just days before the dedication ceremony.

The sparkling new building marked a watershed moment for the region's cheese industry, which had evolved considerably from its humble farmstead roots. To the general public, the facility was simply the new modern face of Tillamook Cheese. But the reality was more complex. In fact, the factory was owned by the four largest area manufacturers—the Holstein, Maple Leaf, Cloverleaf, and Tillamook Creamery plants—all of which had merged their operations into an organization called the Tillamook Cheese Association (later known as the Tillamook Cheese and Dairy Association) and constructed the plant. The familiar cooperative marketing association, the Tillamook County Creamery Association, still existed as a separate legal entity that kept an office at the new plant. Eleven other independent operations still produced cheese in Tillamook County and sold their product as "Tillamook Cheese" under the collective marketing umbrella. All of the new factory's output of cheese and other products was branded and sold as Tillamook under the auspices of the Tillamook County Creamery Association as well. This rather complicated arrangement set the stage for a major conflict that would erupt several decades later.

Perhaps it was residual patriotic fervor that engendered a competitive rivalry among the Northwest's various cheesemaking communities after World War II. In 1945, the *Twin Falls Times-News* announced that the Jerome

Cooperative Creamery cheese plant in Twin Falls, Idaho, was the "largest cheese factory in the West," having produced 3.75 million pounds of cheese the year prior, a good portion of which went to the war effort. The Jerome Co-op took in milk from a number of area dairy cooperatives including the Twin Falls County Cooperative Dairymen's Association and cooperatives in Buhl and Gooding. Not far behind was a plant in Rexburg, Idaho, operated by the Upper Snake River Dairymen's Association, which produced two million pounds of cheese in 1945. Products from both of the Idaho cooperatives were marketed and sold through the Los Angeles-based Challenge Cream and Butter Association.

In a 1947 article, the *Seattle Daily Times* claimed one of its own home state operations as the "largest plant under one roof" in the Pacific Northwest. The paper was careful to distinguish the plant in Battle Ground from rival Tillamook in Oregon by adding that, while Tillamook's collective production level was larger, Tillamook's output originated from multiple facilities. The cheese factory in Battle Ground, a small town in southwest Washington, first opened in 1923 in fertile agricultural territory just north of the Columbia River once occupied by the Hudson's Bay Company. The plant enjoyed almost immediate prosperity; just three years after opening the Battle Ground Dairymen's Union was busily remodeling and expanding its facilities. In 1928 the plant reported that it produced over one million pounds of cheese, enabled in part by adding a night shift. Cheese production continued on an upward trajectory, increasing to nearly two million pounds in 1930. By the 1940s the plant's products were being marketed under the Darigold cooperative umbrella and the operation diversified into producing a variety of products including ice cream, cottage cheese, and condensed milk. In 1946 the plant at Battle Ground produced three million pounds of cheese; that same year the Tillamook County Creamery Association reported a combined output of 9.4 million pounds.

Meanwhile, in 1945, *The Daily Chronicle* in Chehalis, Washington, trumpeted the construction of a new Kraft cheese factory in that city to be opened in partnership with the Lewis Pacific Dairymen's Association. Kraft already had a history in Lewis County; in 1940 it leased a portion of the Lewis Pacific Dairymen's existing dairy production facility and began making cottage cheese there. The new Kraft plant was slated to produce cottage cheese, cream cheese, and processed cheese. While the paper did not directly comment about the Kraft plant's size in relation to other plants, surely the Kraft name spoke for itself. By then Lewis County had built a well-established reputation as a state dairy powerhouse and the Lewis County Dairymen's Association cooperative already operated a large dairy plant in Chehalis, which produced cheese, butter, and powdered milk.

Not to be outdone, Lynden, Washington, boosters claimed that their town possessed the largest cheese factory *in the world*. Lynden, a small town in Whatcom County near the Canadian border, had spawned several cheese factories and creameries during the earliest years of settlement. As early as 1902 a cheese factory at Lynden was producing in excess of nineteen thousand pounds annually. By 1912 there were several creameries in Lynden including the Lynden Creamery and the Banner Creamery, the latter of which expected to open a cheese factory within a few months. The Whatcom County Dairymen's Association, a local cooperative that also operated a plant in Bellingham, opened a cheese factory in Lynden in 1922, and business really took off because the Whatcom County Dairymen's Association was part of United Dairymen, the cooperative marketing organization that adopted the successful Darigold brand. United Dairymen turned the Lynden plant into one of its central milk processing facilities in the state—in fact, the Lynden plant still produces powdered milk products for Darigold today.

Clearly the cultural conversation about cheese was changing. The partisan boasting reflected just how far the Pacific Northwest dairy landscape had evolved by the mid-twentieth century. Decades earlier, small creameries and cheese plants dotted the rural landscape of Oregon, Washington, and Idaho. But production had long since consolidated into increasingly larger processing and manufacturing plants strategically positioned between farms and cities. Next-generation plants like the newly constructed Tillamook facility combined equipment for multiple functions including making cheese, butter, and ice cream, fluid milk bottling, and whey drying under one roof. Flexible functionality allowed producers to accommodate market conditions; where fluid milk brought higher prices, the plant could bottle milk. If cheese or butter prices were higher or the company secured a big contract for these products, the plant could focus on butter and/or cheese. Enabled by these large-capacity plants, production levels of cheese and other dairy products reached multiple millions of pounds annually. On-farm cheese production for commercial sale had essentially disappeared.

The Food Standardization Movement

The rapid expansion of a number of food industries (including the dairy and cheese industries) during the twentieth century did not escape the attention of government regulators. As early as 1906, Congress passed the Pure Food and Drug Act, which among other things created the Bureau of Chemistry within the Department of Agriculture, which later became the Food and Drug Administration (FDA). The legislation was motivated by the proliferation of fraudulent, defective, and unhealthful products on the market that seriously threatened public health and safety. Manufacturers and merchants regularly mixed cheap ingredients with the real thing to gain more profit—common frauds included grinding up soap and adding it to chocolate or mixing plaster of paris with flour. In 1899 in Cincinnati, one thousand people were sickened from eating beef that had been treated with a formaldehyde solution known as freezine, often used to preserve meat and milk products to make them appear fresher than they actually were.

The otherwise pastoral world of dairy products was not exempt from such corrupt practices. During the nineteenth century, short-weight butter and watered-down milk were sold everywhere. Another commonly perpetrated hoax was the making and selling of so-called "filled cheese," a frightening-sounding product made of skim milk combined with lard or other oil fillers, creating what must have been an only marginally edible cheese passed off as the real thing. The filled cheese fraud grew to be a big problem in the industry as cheese factories on the East Coast and in the Midwest began shipping large volumes of bad cheese to Europe during the latter half of the nineteenth century. According to one report, approximately twenty million pounds of filled cheese was produced in the United States annually around the turn of the century. Pacific Northwest producers were not immune to the lure of making filled cheese for profit; in 1909, for example, an inspector discovered that cheese produced at a plant in Newberg, Oregon, contained cottonseed oil. The development of oleomargarine in France and its introduction and manufacture in the United States during the nineteenth century also generated much controversy in the dairy world. Shady manufacturers made cheaper-to-produce oleomargarine and tried to pass it off as real butter, reaping real butter's higher prices.

While the Pure Food and Drug Act gave government authorities significant leverage in fighting the war against fraud and misrepresentation, manufacturers quickly developed ways to skirt the laws. Sophisticated companies circumvented liability by employing invented names such as "Bred Spred," allowing them to argue that because their products did not purport to be anything specific they could not be prosecuted for selling misbranded or adulterated food. The Food, Drug and Cosmetic Act of 1938 expanded the FDA's authority to create legal standards for all manner of food products. The agency went on to develop standard definitions, including detailed ingredient lists and manufacturing recipes, for products as diverse as peanut butter, canned tomatoes, and jam. It was only a matter of time before the FDA turned its attention to dairy products.

Meanwhile, scientists were already devoting considerable time and energy to the issue of dairy products safety. Early research focused on how long various types of pathogenic bacteria could survive in cheese. While results varied with time and temperature, scientists showed that aging decreased the viability of a number of types of harmful pathogens in cheese to at least some degree. Public concern over the possible spread of tuberculosis through meat and milk prompted Professor J. L. Sammis of the University of Wisconsin to conduct experiments using pasteurized milk to make cheese in 1912. Pasteurization, a process developed by Louis Pasteur in France in the nineteenth century, involves heating milk to a prescribed temperature in order to destroy harmful microorganisms. But during this period the cheese industry was mixed on the relative merits of pasteurized milk versus raw milk. The prospect of industry-wide pasteurization of milk used for making cheese represented a significant paradigm shift to which many in the industry were resistant. Some experts felt that pasteurizing milk diminished the flavor and quality of cheese, though others pointed out that pasteurized milk made for more consistent quality cheeses. At least one scientist feared that pasteurization of raw milk used for cheese could itself become a tool of fraud, effectively covering up all manner of unsanitary practices on farms and in manufacturing plants. Other dairy industry scientists worried about the potential costs of obtaining and operating pasteurizing equipment and the additional burden of further regulatory control of milk and milk products going forward.

The trend toward regulation of cheese gained considerable momentum during the 1930s and '40s, primarily because of a steady increase in outbreaks of food-borne illness attributed to cheese. Between 1935 and 1944, 824 illnesses and eighteen deaths in the United States were traced to a variety of cheese products including cheddar cheese, cottage cheese, and cheese curds. In 1944, California passed the first law in the nation requiring raw milk cheese to be aged for sixty days, motivated by an outbreak of typhoid fever in California and Nevada that was traced to cheese and in which seventy-nine people were sickened. Colorado passed a law in 1945 that required a curing period for cheese of one hundred twenty days. By 1946 eight states had either passed or were reviewing legislation requiring a mandatory curing period and a variety of labeling requirements for cheeses made with raw milk. So pervasive was the trend that an increasing number of cheesemakers in other states including Missouri and Washington began using pasteurized milk as a matter of course even though their state laws did not specifically require it.

FDA action was inevitable not only because scientific and legislative momentum was heading toward the regulation of cheese made from raw milk, but because the variety of laws appearing on the books in different states presented a tangled web that threatened to significantly complicate interstate commerce. In 1949, the FDA codified the building momentum when it created legal definitions for various types of cheese. Among the most significant aspects of the new standards was the requirement that most varieties made with raw milk be aged for at least sixty days at a minimum of 35 degrees Fahrenheit before being sold. While these regulations weren't controversial at the time, they would have significant implications for cheesemakers in future decades.

Rindless Cheese—An Industry Revolution

> Where is the bright inventive genius who will invent the rindless cheese? Is he yet born? Surely if he is not in existence he ought to be, for the times demand it.
> *Elgin Dairy Report*, October 6, 1913

In the early twentieth century the idea of a rindless cheese was no more than a distant fantasy. Traditionally, cheese manufacturers made and aged cheese in variously sized loaves and wheels that were shipped to wholesalers and then to shops, which cut and wrapped cheese by hand to order for consumers. But these practices clearly had disadvantages. Cheese manufacturers did not like the fact that aging cheese shrank over time due to moisture loss, reducing profits. In addition, storage warehouses were breeding grounds for all manner of defects including mold and other forms of environment-induced deterioration. Consumers were unhappy because when they purchased cheese they often ended up with a hunk of inedible rind along with their cut piece of cheese; they felt cheated out of their hard-earned money.

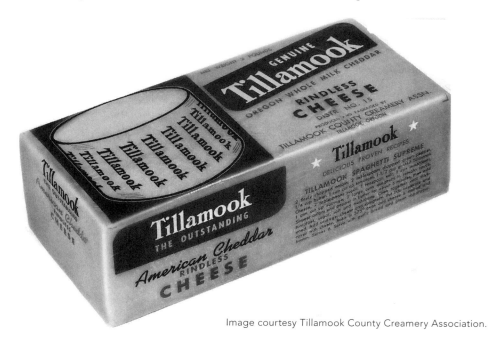

Image courtesy Tillamook County Creamery Association.

By the mid-twentieth century cheese manufacturers and dairy scientists had been searching for decades for ways to solve these problems. Wax had been one of the most commonly employed solutions to protect an aging cheese; though generally successful, waxing was prone to problems, especially mold growth and other types of spoilage, which often took root in cracked wax or in crevices between the wax and the surface of the cheese. On the manufacturing side, wax was often misapplied and was prone to breaking as it was handled during shipping. Over the years industry analysts experimented with a variety of alternate protective coverings for cheeses such as tin cans, aluminum foil, Parakote laminate products, and Patapar vegetable parchment, but none proved to have real staying power or widespread commercial appeal.

We have the Goodyear Tire & Rubber Company to thank for developing Pliofilm, a substance that revolutionized the food industry. The clear plastic-like film, actually a rubber derivative, was invented in 1934 and used during World War II for packaging of airplane engines and other delicate precision military instruments. Troops also used the material to protect weapons and supplies against the elements. Pliofilm entered the world of civilian applications in the 1940s and Goodyear heavily promoted its product in the media of the period. "Here's Why Food Should be Pliofilm Protected," they shouted from the rooftops, touting among other things its ability to prevent dehydration in the refrigerator, thus insuring freshness and quality.

Pliofilm was a breakthrough product that ushered in a new era of merchandising and manufacturing in the cheese industry. Pliofilm prevented the moisture loss typical of aging cheese, enabling the elimination of a rind. Rindless aging also prevented the invasion of molds and other unsightly and/or harmful microorganisms both at the aging stage and in the final packaged retail product. Pliofilm was also appealing to manufacturers because it was cheaper, simpler, and much more attractive to shoppers than that other experimental cheese packaging technology of the time: metal cans. The packaging allowed cheese companies to target their brands directly to consumers and to create demand based on that appeal. Manufacturers heavily promoted the novelty product with ads that proclaimed the virtues of rindless cheese: "Menus Brightened Easily, Economically with Darigold Natural Rindless Cheese—No Rind, No Waste" and "Natural Rindless Cheese Delights Housewives." Among its other virtues, Pliofilm enabled manufacturers to be creative: Tillamook started selling a product it called Club Cheese, a flavored spread popular with manufacturers of the period because it was made from trim ends and pieces of cheese that would be otherwise discarded, and wrapped in an attractive Pliofilm wrapper.

In the 1930s researchers at the University of Wisconsin studied ways to improve the time- and labor-intensive method of selling cheese cut to order over the counter. Among their experiments was one in which they wrapped small pieces of cut cheese and placed the pieces on a store counter. One participating retailer saw a 400 percent jump in cheese sales. Pliofilm allowed cheese manufacturers to develop consumer-friendly grab-and-go sizes and allowed the retailer to eliminate the cheese counter, all while increasing cheese sales and consumption. "There's been an innovation in the packaging of natural cheddar cheese that may make the familiar round wheels of store cheese as obsolete as their former accompaniment, the cracker barrel," noted the *New York Times* in 1948—and that was exactly what happened.

Smaller-scale Cheese Factories

Despite the dominance of large-scale cheese production in the Pacific Northwest during the 1940s and '50s, smaller regional producers found ways to survive in the post-World War II mass production landscape. Many did so by exploiting an emerging market niche for specialty products such as European-style cheeses, which they could produce and sell at a lower price than imports. Others such as Rogue River Valley Creamery in southern Oregon carved out a niche selling their output to national-scale companies like Kraft or Borden. Despite being small by contemporary industry standards, these producers were still sizeable factories considerably larger than their pioneer farmstead counterparts, with production levels generally in the range of at least one million pounds of cheese per year.

Mt. Angel, a small town in Oregon's Willamette Valley, east of Salem, grew in part because Swiss monks of the Benedictine order chose it as the site of a monastery in 1881; a convent was later founded there as well. The Mt. Angel Cooperative Creamery, formed by area dairy farmers in 1912, eventually became well known across the region for its Rose Valley brand butter and cheddar cheese. The creamery's production was diverse and included, at various times, butter, cheese, ice cream, buttermilk, and powdered milk. For a period during the 1930s, the Mt. Angel Cooperative sold most of its output of butter directly to Safeway retail grocery stores, a sales coup that no doubt went a long way toward sustaining the company (and local dairy farmers) through the Great Depression. Though the creamery at Mt. Angel closed in 1969, Mt. Angel brand cheddar was produced and sold by Darigold through the 1970s and its Rose Valley brand butter continues to be produced by Farmers Cooperative Creamery in McMinnville, Oregon.

Some of the era's cheesemakers found success exploiting consumer interest in European-style cheeses. In 1932, brothers Beck, Jess John, and

Chris Hansen, along with partner C. J. Christensen, established a small cheese factory in Nisqually, Washington. south of Tacoma. The Hansens, originally from Denmark, developed a product they called Danish Banquet Cheese, a soft-cured brick cheese made by what they claimed at the time was a secret process, but seems to have resembled Edam. The product took off almost immediately; within one year, production reportedly jumped from one hundred fifty to twelve hundred pounds of cheese per day and the brothers constructed a new plant to accommodate the growth. In 1951, the Danish Banquet Cheese Co. opened a retail facility south of Tacoma. The deli-style store featured "modern fluorescent lighting, a large walk-in refrigerator holding approximately 30,000 lbs of cheese and refrigerated display show cases." The so-called Cheese House carried upwards of a hundred varieties of cheeses both foreign and domestic, cured meats, and self-described obscure Scandinavian specialties, as well as coffee, sandwiches, and ice cream. Cheeses from the Danish Banquet Cheese Company were sold in the Northwest through the 1950s. The Hansens eventually sold their production operation and it became the Olympia Cheese Company, which made cheese through the 1980s.

Carl Hopperdietzel founded an eponymous cheese factory in St. Anthony, Idaho, near Idaho Falls in 1952. Hopperdietzel, a Wisconsin native, trained in the craft of making cheese in his home state and operated a cheese factory there during the 1920s. The Hopperdietzel family then moved west, travelling first to Colorado and then Idaho in the 1930s. Carl Hopperdietzel worked for a number of large factories in Idaho including the Lower Snake River Cooperative before starting his own plant in St. Anthony. The plant produced an average of 1.6 million pounds of monterey jack and American cheeses annually. Hopperdietzel's survival strategy was simple: he sold all of his plant's output to industry giant Kraft, whose processed cheese plant was ninety miles away at Pocatello, reserving just a small amount for over-the-counter sales at the factory.

Farther north in Sandpoint, Idaho, William Beyer started the Pend Oreille Cheese Co. in 1960 in an old meat packing plant. Beyer, who ran a farm and cheese factory in Argyle, Wisconsin, prior to coming to Idaho, came to the Pacific Northwest after seeing an ad for a cheesemaker in a trade publication. He soon noticed that many Idaho dairy farmers were dumping milk that processors would not purchase due to artificial price supports. Seeing that there was plenty of available milk, Beyer took advantage of the situation and started his own cheese factory. By the late 1960s the factory had twenty-eight employees and was shipping a train carload of five-hundred-pound barrels of cheese every ten days to Utah for purchase by the federal government. Pend Orielle later diversified into producing a variety of consumer-friendly cheeses such as its alpine-style Mountain Gem. The operation closed in the late 1990s and Idaho-based Litehouse Foods currently produces blue cheese in the same building in downtown Sandpoint.

The southern Oregon coast has a long history of dairying and cheese-making stretching back to the days when the first settlers made their way to the region, but by mid-twentieth-century just a few cheese factories were left. The Superior Cheese Co. operated a sizeable factory at Myrtle Point that produced cheese exclusively for the Safeway Corporation during the 1950s. The cheese factory at Reedsport, owned by Danish cheesemaker Sven Knudsen, produced cheddar cheese for Von's Grocery Stores in southern California from the 1950s to the 1970s. "Made by Sven Knudsen," read one ad, which further elaborated that the cheese was made "Where Lush Grass and the Umpqua River Meet the Blue Pacific." The Reedsport Factory closed in 1977.

Nearby in Bandon, the Coquille Valley Dairy Cooperative operated the local cheese factory, having taken over the operations of the Bandon Cheese and Produce Co. after a devastating fire swept through the town and destroyed the original factory in 1936. By the 1970s the co-op was producing over two million pounds of cheese annually, mostly cheddar along with some monterey jack. But the Bandon factory began to struggle during the 1980s and even closed for a period of time in 1987. "The price of milk [has become] so high that we couldn't produce cheese and sell it profitably," said Michael D. Jones, general manager of the Bandon plant, at the time. A number of buyers including the Tillamook County Creamery Association entertained the idea of purchasing the Bandon plant; Italian company DeGrassi Foods, which had been considering building a cheese plant in Eugene, nearly purchased the Bandon operation, but plans fell through.

In 1988, the Bandon factory was resurrected by a group of local buyers that included former head cheesemaker Bob Howard, and this group restarted the plant under the name Bandon Foods. The factory changed hands again in 1991 when local dairyman Joe Sinko and a partner purchased the creamery. The plant was able to continue producing cheese at least in part because it provided and packaged cheese for Harry and David Corporation, a mail-order gift retailer based in Medford, Oregon. In 2000, the Tillamook County Creamery Association purchased the factory, but closed it for good in 2002.

Goat's milk cheese production continued in the Pacific Northwest through the 1950s, though it did not maintain the momentum generated during the World War II era. Consumers who may have turned to goat's milk and/or goat's milk cheese during the war years because the products were ration-free apparently did not continue consuming goat cheese after the war; clearly it would take more than just cost or convenience to win over the average

American consumer to the idea of consuming less familiar styles of goat's milk cheeses. Northeast of Seattle in Gold Bar, Washington, Bearga Goat Dairy was one of a handful of remaining farmstead cheesemakers that offered goat's milk cheese for sale through classified ads in the *Seattle Times*. The Wallace family, owners of Mystic Lake Goat Dairy in Redmond, Washington, started selling raw goat's milk in 1961. The dairy rolled out a Bulgarian-style goat's milk yogurt in 1970 and the operation grew quickly, distributing its milk and yogurt to health food shops and restaurants along the West Coast and Canada. The ever-increasing popularity of Mystic Lake's products signaled that a new era was on the horizon.

M. P. Eggers once referred to himself as the "Henry Ford of the goat cheese business," and perhaps he wasn't far from the truth. The son of German immigrants, Melvin Eggers grew up working in his family's commercial fish business in Tacoma but spent the rest of his life among goats. As we already saw in chapter three, Eggers had started out as a farmstead operation, selling the milk from his own goats at a farm in North Bend, Washington, but once he moved south to Chehalis in 1944 he began purchasing milk from a number of small goat dairies scattered throughout the area and used the milk to make cheese. Eggers' Briar Hills Dairy, in business since the 1920s, produced goat's milk cheese and whey products out of a small plant in Chehalis, Washington. By the 1950s two major fires had destroyed Eggers' facilities and equipment but he still managed to keep going. An inveterate individualist, Eggers once attributed his success to "making his own brand of cheese that could not be duplicated." His product line varied over the years but included a swiss-style goat's milk cheese as well as a fresh cheese Eggers called "Pliny's Delight," which he said he developed through reading the writings of Pliny, an ancient Roman philosopher. Another cheese called Cascadian was aged about six months. His well-entrenched mail order business, developed over many

Peter Eggers making goat's milk cheese at Briar Hills Creamery in Chehalis, Washington, 1973. Photo courtesy Bill Moomau.

decades, likely kept him afloat over the years—Briar Hills was reportedly one of the heaviest parcel postal users in the Chehalis area.

Son Peter took over the Briar Hills operation when Melvin Eggers died in 1968. By then interest in goat's milk products was on the rise, and Briar Hills cheeses began to be sold at mainstream outlets such as Puget Consumer's Co-op in Seattle and were bought by a new generation of mail order customers. Peter Eggers sold the business in 1982 and the operation changed hands several times. Current owner Frank Stout phased out the cheese side of the business but continues to produce goat's milk whey products including Whex under the Mt. Capra brand in the same building in Chehalis.

Swiss Cheese in the Pacific Northwest

Swiss immigrants brought their cheesemaking knowledge and technology with them when they immigrated to the United States during the late nineteenth and early twentieth century during periods of economic instability and political upheaval in Europe. Prior to World War II dozens of independent swiss cheese factories came and went across the Pacific Northwest, among them Fred Zobrist's factory in the Whatcom County town of Acme, Washington; the Ernst Swiss Cheese Factory in North Plains, Oregon, west of Portland (which burned down just a year after opening in 1920); and the Stanfield Swiss Cheese Factory in Umatilla County in eastern Oregon.

Still, swiss cheese production lagged significantly behind cheddar in the United States during the early twentieth century. "The fundamental problems involved in the manufacture of Swiss cheese are but little understood," noted government scientists at the USDA's Division of Animal Industry in 1907. "[T]he whole process seems to be 'rule of thumb' rather than any scientific method." Even today, scientists consider swiss cheese to be one of the more challenging styles of cheese to make due to the complex microflora involved and the many technical steps involved in its manufacture. Swiss cheese posed physical challenges as well. Traditionally, swiss cheese (known as Emmenthaler in its home country) was made in large wheels weighing in excess of two hundred pounds. The wheels of cheese were washed in salt water during the early stages of the maturation process; as if that weren't challenging enough, each wheel required carefully controlled aging typically lasting a year or more. "The swiss cheese factory requires a special curing room, usually four to five times as big as the make room," noted one industry analyst in the 1920s. "This is commonly built up into the side of a hill, or else earth is banked up on both sides of the building up to the eaves, so as to keep the temperature constant." Once properly matured, the finished product—an enormous wheel of cheese—proved unwieldy for wholesalers and retailers to handle and merchandise. As a result of these challenges, swiss cheese production in the United States first evolved as a specialty niche that those with the technical skill and know-how wasted no time exploiting. Swiss immigrant communities in Wisconsin and Ohio in particular became thriving centers of swiss-style cheese production in the United States.

Brothers Ernest, Paul, and Fred Brog were among the early pioneers of swiss cheese production in the western United States. Born in Berne, Switzerland, home of Emmenthaler production, all three learned the cheesemaking trade there before emigrating to the United States. Each initially plied his trade in Monroe, Wisconsin, then one of the centers of swiss cheese production in the nation, before heading west. Ernest Brog was the first to seek greener pastures; he ran several cheesemaking operations in Wyoming and Idaho before finally settling just east of the Idaho border. His Star Valley Swiss Cheese Co. in Freedom, Wyoming, began producing swiss cheese in 1927. Brother Paul Brog arrived next in 1930 and took over a cheese factory at nearby Thayne, Wyoming. Paul later became manager of the Bear Lake Valley Cooperative just to the west in Paris, Idaho, and eventually moved there. Fred Brog came west last in 1931 and took over management of the Upper Star Valley Swiss Cheese Association in Afton, Wyoming. The Brog family's burgeoning cheese empire, which a contemporary trade publication termed "Switzerland in the Rockies," was responsible for producing a combined three million pounds of swiss cheese per year during the 1940s.

The Brogs helped establish a regional pocket of swiss cheese production in the Rocky Mountains that included a number of small valley towns in eastern Idaho, western Wyoming and northern Utah. Among the many factories were those established in Gray's Lake and Grace, Idaho, located about fifty miles apart near the Wyoming border. Grace had been the site of one of the Laabs family's Kraft-sponsored cheese plants during the 1920s; in 1933 brothers Alvin and Melvin Nielsen built the Gem Valley Swiss Cheese Factory there. Not far away in Wayan, Idaho, the Gray's Lake Swiss Cheese Factory started in 1934 with Ernest Brog providing advice and assistance to local dairy farmers. By the mid-1930s this region was responsible for 25 percent of all swiss cheese production in the United States.

After World War II the Brogs consolidated their Wyoming operations into one large factory located in Thayne. The Star Valley Cheese Factory was billed at the time as the largest swiss cheese factory in the world. In 1944 Paul Brog purchased the Rocky Mountain Creamery in Salmon, Idaho (which later became the Salmon Valley Cheese Co.), where he produced swiss cheese separately from the Wyoming operation. By then, however, the Brogs were facing significant competition from a large swiss cheese plant one hundred

Swiss Village Cheese Co., Nampa, Idaho

In 1973, Idaho cheese entrepreneur Paul Brog Jr., son of Paul Brog of the Swiss Brog brothers, joined forces with Ed Gossner Sr. and his son Ed Gossner Jr. of Utah's Gossner Foods to found the Swiss Village Cheese Co. in Nampa, Idaho. The founding of the plant represented the coming together of two Swiss cheesemaking families, each with a long history in the region's dairy industry.

The Swiss Village factory started out with just five employees. Within a few years it employed fifty people working around the clock producing twelve varieties of cheese. By 1976 the plant was producing about twenty-six thousand pounds of cheese daily. The factory became a popular tourist destination for travelers along the freeway between nearby Boise and points east and west. Swiss Village worked hard to connect with tourists and the local community by offering viewing and tours, and of course, fresh cheese curds. A restaurant was added later and cheese brogies (deep-fried cheese curds) were a menu favorite. The retail shop sold cheese from Brog's Salmon Valley plant including the popular Idaina as well as cheese made at the Gossners' plant in Utah, and that made in-house. The factory became a popular showcase for the local dairy community.

In 1990, the J. R. Simplot Company, well known in the Idaho potato industry, purchased the Gossners' interest in the plant as part of an effort to branch out into the dairying business. The next year, Simplot also purchased a cheese plant in Arpin, Wisconsin, as well as the Washington Cheese Company in Mt. Vernon, Washington. Simplot expanded the Swiss Village plant significantly, then sold the operation to Sorrento Lactalis in 1999. Sorrento Lactalis, owned by France-based Groupe Lactalis, continues to operate the plant today, producing Italian-style cheeses including mozzarella, mascarpone, and string cheese. The plant is no longer open to visitors.

miles to the southwest in Smithfield, Utah, operated by the Cache Valley Dairy Association (which claimed to operate the largest swiss cheese plant in the world). There, Swiss-born cheesemaker Ed Gossner supervised the production of multiple million pounds of swiss cheese, all large wheels.

Meanwhile, scientists were hard at work creating techniques that made swiss cheese more amenable to large-scale commercial production. In particular, the development of a process for making rindless swiss cheese by Kraft researchers in the 1940s changed the industry significantly. These developments began to put pressure on the boutique market that the Brogs and producers in the Midwest had been profitably exploiting for decades. By the 1950s the Star Valley Cheese Factory began to diversify its production into more lucrative lines of cheese such as mozzarella and provolone that were growing in popularity. Smaller plants such as those in Grace and Gray's Lake, Idaho, struggled and eventually closed. California-based SV Cheese Co. purchased the Star Valley factory in 1993, saving it from closure, but the plant finally closed for good in 2005. Paul Brog's son, Paul Jr., took over the Salmon Valley, Idaho, plant from his father and branched out into additional varieties of cheese including cheddar and monterey jack, as well as sideline products like candy. Paul Brog Jr. also developed a new cheese, a proprietary style he called Idaina, a hybrid danish-swiss-style cheese with holes that the company sold in a range of stages from a milder, younger form to a sharper cheese aged as long as four years.

In 2010, United States manufacturers produced over three hundred million pounds of swiss cheese. Just one cheesemaker in the United States, Edelweiss Creamery in Wisconsin, still produces swiss cheese in traditional large wheels.

Italian Cheese in Washington

During the early twentieth century the largest population of Italian immigrants on the West Coast resided in San Francisco, California, but many Italians also flocked to the Pacific Northwest. Like other Europeans coming to the region, they worked in the region's rapidly expanding industries, laboring in railroad yards, lumber mills, mines, and factories. Some Italians became farmers, tending truck gardens on the outskirts of Portland, in the Parkrose neighborhood, as well as in Rainier Valley (once known as "Garlic Gulch"), south of downtown Seattle. Portland's Italian immigrant community established an Italian Market on the east side of the Willamette River that predated the Public Farmers Market.

For years, small markets in these ethnic enclaves sold imported Italian cured meats, cheeses, and other specialties that catered to the tastes of the community. But after World War II, the popularity of Italian foods of all kinds skyrocketed, a rise in interest generally attributed to the expanded palates and perspectives of thousands of returning GIs who came back with an appreciation for Italian foods, including Neapolitan pizza, they'd acquired while fighting in Europe. Pizza arrived on the American culinary scene at precisely the right moment, at least from the perspective of the cheese industry, as it was a vehicle for increasing consumption of mozzarella cheese. The dairy industry wasted no time in promoting this new specialty cheese, as it was eager to patch the postwar production slump. Cutting-edge cheese packaging technologies like Pliofilm and later generation plastic wraps also contributed to the growth of mozzarella sales, as these wrappings enabled the preservation and widespread distribution of fresher, less aged styles of cheese such as mozzarella.

Green River Cheese Co. (Castrilli Family)

During the first decade of the twentieth century, hundreds of Italians emigrated to the Skagit Valley of northwest Washington, attracted by the region's fishing industry and many lumber and railroad industry jobs. Among the Italians who flocked to the region were Louis and Elizabeth Castrilli. Louis Castrilli's father taught him to make cheese in Italy and Louis continued the family cheesemaking tradition once he arrived in the Pacific Northwest, starting a cheese factory in Hamilton, Washington, a remote mining and lumber town in the foothills of the North Cascade Mountains. His sons John and Pasquale (Pat) learned the cheesemaking trade from their father and went on to start the Imperial Cheese Company, which operated several cheese plants in the Skagit Valley including one at East Stanwood and another in Everson near the Canadian border. In 1936, Imperial purchased two additional operations in Lewis County, Washington, the Central Cheese Factory in Chehalis and another at Morton.

Though the Depression took its toll on the Imperial Cheese Co., Pat Castrilli went on to start the Seattle Cheese Co., located on Rainier Avenue in Seattle, in 1939. The plant made about twenty variety of cheeses including mozzarella, provolone, cheddar, and a jack-style breakfast cheese that sold very well as it was a soft cheese that cost only a few ration points, a significant sales incentive during the war. Castrilli sold the Seattle Cheese Co. in 1947 and purchased the Mt. Vernon Butter Store at the Pike Place Market, a small shop that sold cheese as well as butter, eggs, and canned goods. The Castrilli family operated the store through the mid-1960s.

Pat Castrilli was not out of the cheesemaking business for long. In 1953, he opened the Green River Cheese Co. in Kent, Washington, in an old barn located along the Green River. Like the Seattle Cheese Company, this plant turned out a variety of cheeses including mozzarella, monterey jack and dry jack, and other Italian specialties. A good portion of the company's output was produced under contract and labeled for regional grocery store chains like Albertson's, Big Bear, and Dunham's. Ten years later the plant moved to a larger facility, also in Kent, where it continued producing specialty cheeses and also

Louis Castrilli at his Hamilton Cheese Factory in Hamilton, Washington. Photo courtesy Skagit County Historical Museum.

branched out into the wholesale business, stocking products like tomato sauce and jalapeno peppers that it sold to pizza parlors and Mexican restaurants. The plant closed in 1994.

Mazza Cheese Co. (Mazza Family)

The small town of Orting, Washington, lies tucked in the hills east of Tacoma in the shadow of Mt. Rainier. Charles Mazza Sr., originally from Naples, Italy, purchased the town's cheese factory in 1929. Mazza specialized in Italian-style cheeses and sold the bulk of the factory's output to Italian grocery stores and to the Puget Sound's growing immigrant population.

Charles' son Louis took over in 1934 and the operation continued to thrive. By the 1950s and '60s the public's tastes had evolved and the once obscure specialty Italian cheeses became increasingly profitable. By 1963, Mazza Co. was making over one million pounds of mozzarella, jack, and caciovallo cheeses annually and distribution of its Seal brand broadened to retail grocery markets all over the Pacific Northwest.

In 1989, CEO Darlyne Mazza, who took over operations after her father-in-law Louis passed away, announced plans for a $20,000,000 expansion into a new facility in nearby Sumner, Washington. The company expected to triple its capacity and develop a side business selling the plant's dried whey by-product. In 1991, the company was sold to corporate giant Beatrice Foods; news reports of the period indicated that Mazza Cheese Company was on the brink of bankruptcy. Beatrice closed the Sumner plant for good in 1998.

Tempest in Tillamook

1949 had been an important year for Tillamook County's dairy industry. The newly built cheese factory on Highway 101 drew national attention to the area's thriving cheesemaking industry. Yet despite all the enthusiasm generated by its opening, the cooperative nearly imploded from within just a few years later. The split was a bitter and divisive chapter in the history of both the surrounding community and the organization itself.

The seeds of the dispute had been sown in the 1940s when the new plant was built. The large plant, owned by the Tillamook Dairy and Cheese Association (TCDA), produced over half of the cooperative's collective output of cheese and thus wielded more power than any of the other plants under the Tillamook County Creamery Association (TCCA) marketing umbrella. According to TCCA manager Beale Dixon, "[t]here was always a bit of a rivalry you might say between the small plants and the big plant . . . although the big plant was built originally with the hope that [the small plants would come into the big plant]." Still, few would have guessed that the differences would nearly tear the organization apart.

In October of 1962, just a day before the infamous Columbus Day Storm, TCDA filed a lawsuit against the TCCA. The furious tropical storm, which generated winds in excess of 150 mph and caused hundreds of millions of dollars in damages along the entire West Coast, was a fitting metaphor for the years of mayhem that were about to descend on the city and county of Tillamook. At its core, the issue was control; in the suit TCDA accused TCCA of attempting to control its business practices and asked the court to determine the rights of each party. The most significant aspect of the judgment, rendered several months later, was that TCCA, a cooperative marketing agent, could not legally dictate the independent TCDA's activities. The judgment opened the door for TCDA, and it went on to do the unthinkable—it split from the TCCA altogether.

Litigation flew back and forth for years on both the state and federal level. In 1964, TCDA asked a court to appoint a receiver to run the TCCA, accusing it of violating the cooperative agreement by selling cheese made in Minnesota. The next year, TCDA filed an antitrust suit against the TCCA, accusing it of restraining TCDA's trade and seeking $7.7 million in damages. TCCA responded by filing a countersuit asking for $20 million in damages. Meanwhile, TCDA hired a Portland marketing firm that came up with the idea of selling TCDA cheese as "Premium Brand" Tillamook Cheese. This led to yet another lawsuit involving which organization had the right to use the brand name "Tillamook," which over the decades had gained considerable market traction. TCDA won a significant legal victory in 1965 when the 9th Circuit Court of Appeals upheld its right to use the Tillamook name to brand its own separate line of cheese and dairy products.

The substance of the dispute was rooted in contemporary milk policy. During the early twentieth century, in the wake of the milk safety era, most states adopted some form of a fluid milk grading system. Grade A milk, of the highest quality, was considered fit for human consumption and lower grades were classified as acceptable for other uses, including the manufacturing of cheese or butter. Farmers who produced Grade A milk received a higher price for their milk, but were also expected to maintain much higher sanitary standards than farmers producing lower grades. While many farmers began converting to Grade A during this period because of the financial incentives, the grading policy had the ancillary effect of creating classes of farmers, a fact that became especially significant in a cheese-producing area like Tillamook County.

During the 1950s, prices for fluid milk were high and the TCCA, under Beale Dixon, developed a profitable side business shipping Grade A fluid milk to the Portland marketplace. Dixon viewed this move as good for the organization because fluid milk brought in more money than cheese at the time. But dairy farmers in other parts of Oregon began to resent Tillamook's advances into the milk marketplace. At the time, Oregon law controlled the number of Grade A dairies as a means of regulating the milk market, promoting quality, and stabilizing prices. TCCA head Dixon vehemently disagreed with the state's Grade A milk policy, believing that anyone who produced milk to Grade A standards should be able to sell it for the higher Grade A price. From the perspective of some Oregon dairy farmers, Dixon's actions threatened to bring down the whole system; a substantial portion of TCDA farmers agreed with their brethren. "But really, the problem boiled down to a disagreement about marketing," says Floyd Bodyfelt, Tillamook native and former professor in the Food Science and Technology Department at Oregon State University. "Some farmers in Tillamook wanted to keep putting all of their milk into cheese. But Dixon saw that cheese prices constantly fluctuated and so he cultivated the market for fluid milk." And the more milk that TCCA sent out to the fluid milk market, the less milk was available to make cheese at the large TCDA factory. That set the stage for the long and protracted conflict between the cooperative and TCDA, by far its largest and most powerful member.

The dispute grew to epic proportions and members on both sides became increasingly entrenched in their positions. Conflict raged throughout most of the 1960s, pitting neighbors and family members against one another and generating considerable bitterness in the local community. TCDA member Ferdinand Becker received a threatening letter, published in the *Tillamook Headlight-Herald* newspaper, that read in part: "You are to quit and desist from meddling in all affairs pertaining to the TCCA and also TCDA if you don't I shall burn your farm buildings and if that don't stop you I shall get my scope sighted deer rifle and pick you off." TCCA head Beale Dixon was himself a source of controversy, as some felt that he bred divisiveness within the organization due to his forceful personality. A building undercurrent of resentment became starkly evident when Dixon's home and car were damaged by a bomb blast that left a two-foot long crater in his driveway. On another occasion, a man was said to have left the TCCA boardroom during a meeting, returned with a drawer from a file cabinet, and threw it at Dixon. Perhaps in a bid to garner local support to its cause, TCDA ran numerous ads in the local paper promoting its milk, cheese, and ice cream, with slogans clearly designed to appeal to the local populace such as "Tillamook's Own Dairy Products for Tillamook People."

After severing ties with the TCCA, TCDA found itself in the precarious position of marketing its own products, including managing the entire process of shepherding its cheese from its factory to store shelves. This was no simple task in a highly competitive marketplace where Tillamook was a relatively small player compared to far larger national brands like Kraft. TCDA struggled with this from the outset and was even forced to temporarily shut down its plant at one point in 1965 because of a reported reduced volume of business. While TCDA was no doubt banking on the already solid reputation of the Tillamook name on the West Coast, the difficulties inherent in distinguishing its "Premium Brand" from what was simultaneously being produced and marketed as "Tillamook" brand dairy products by TCCA worked to the detriment of both organizations. News reports of the day indicated that TCCA threatened legal action against its wholesale dealers if they bought any products from rival TCDA, a move that must have discouraged wholesalers and retailers from selling *anything* coming out of Tillamook.

TCDA eventually entered into a contract with California-based Foremost to market its cheese and dairy products, though it parted ways with the company less than two years later. In mid-1967, TCDA entered into an agreement with a large dairy cooperative based in Arlington, Washington, for one hundred thousand gallons of milk daily, enough to enable TCDA to triple its output. "It is expected that this added production will restore [TCDA's]

Ladino Cheese Factory Purchased by Cult

Swiss native Adolph Woodrich lived in Idaho and Montana before moving to Eagle Point, Oregon and opening the Ladino Cheese Factory in a former granary building in 1930. Ladino is the name of a common variety of clover, so the name would have conjured up pastoral images of cows grazing on lush green pastures, giving sweet milk that made delicious cheese. Cheesemaker Woodrich became popular with local children because he would hand out warm cheese curds if they came by the factory at the right time.

Meanwhile in Los Angeles, a man named Arthur Bell published a book called *Mankind United* in 1934. The book spun an elaborate conspiracy theory involving a cadre of Hidden Rulers that controlled society and a group of good, moral people called the Sponsors that had formed a secret society to combat them. Bell positioned himself as the conduit through which outside adherents could join the Sponsors' movement and gained a considerable following. Mankind United got into trouble during the late '30s and early '40s with its vocal denunciation of the war effort and for, among other things, promoting the idea that the Pearl Harbor tragedy was engineered by the U.S. government. Bell amassed a fortune estimated to be worth over $4,000,000 that included property and a variety of businesses including restaurants, canneries and machine shops up and down the West Coast, all donated by adherents. *Time* aptly called Arthur Bell "Profit's Prophet."

Under investigation by the FBI and the California Attorney General's office, Bell changed the organization's name to Christ's Church of the Golden Rule and became a tax-exempt religious organization. Christ's Church sank a substantial amount of its cash into property in Southern Oregon, purchasing the Ladino Cheese Factory in Eagle Point along with a fish hatchery, a gladiolus bulb farm, and the Grants Pass Hotel. But by the time of its move into Southern Oregon Christ's Church of the Golden Rule was already in full retreat mode. The Church declared bankruptcy by the end of 1945 and after years of protracted litigation most of its assets, including the Oregon properties, were sold off. A few remaining members persevered, however, and in 1962 they purchased the former Ridgewood Ranch in Mendocino County, California, where what remains of the Church continues to operate today. The former Ladino Cheese Factory in Eagle Point currently houses an antique store.

large factory to the full operating capacity in enjoyed in 1963," said TCDA board chairman Hans Leuthold. Around this time ads began to appear in the *Tillamook Headlight-Herald* touting TCDA-produced monterey jack, a cheese that is cheaper and simpler to make than cheddar. In an odd way, all of that monterey jack cheese may have been a signal that something was not well with the TCDA.

There were several unexpected twists in the long-running melodrama. In the summer of 1967, TCCA was forced to issue a recall when two elderly sisters in Seattle became ill with a staphylococcus infection, ostensibly from eating TCCA cheese. Though health authorities later declared TCCA cheddar free of contamination, the incident could have significantly damaged its reputation and tipped the balance of the conflict in TCDA's favor. But what was perhaps the final blow came from an unexpected source—environmental regulators. During 1967, the Oregon State Sanitary Authority, precursor to the present-day Oregon Department of Environmental Quality, began investigating the dumping of waste into Tillamook Bay. The area's cheese factories, particularly TCDA's large plant, were the primary culprits. The prospect of enormous costs associated with developing and constructing a waste treatment system, on top of the large sums of money it had already spent on litigation and the huge financial commitment recently made to increase its output, was probably the final nail in the coffin for the TCDA.

In September of 1968, the TCCA and TCDA announced that the long-running dispute between the two companies had ended. TCCA acquired TCDA and all of its assets and TCDA was dissolved. All pending litigation was dismissed. Most significantly, all remaining independent cheese plants in the area were subsumed under the TCCA corporate umbrella and, after over fifty years, Tillamook became a wholly united cooperative. Just like that, the fight was over. Beale Dixon remained head of the organization until he retired in 1975. Some area farmers credited Dixon with bold reform, while others refused to sell their milk to TCCA for decades afterward.

By the mid-twentieth century industrial-level cheese production dominated the industry in the United States. In the Pacific Northwest, the cheese industry had come a long way from the days when settlers stepped out of their covered wagons and commenced milking cattle, then traded butter and cheese with the neighbors. The region had grown into a strong center of dairy and cheese production, led by the growth of farmer-owned marketing cooperatives. Through the collective pooling of milk from dozens, even hundreds, of dairy farmers, cheese production levels reached heights that would have stunned even the most optimistic farmstead cheesemakers of decades past. At the same time, consumers across the nation could not get enough cheese. Demand increased steadily, fueled by newly developed merchandising methods and colorful, attractive Pliofilm and plastic packaging. Those smaller producers that managed to survive during this period did so by producing products under contract for national brands like Kraft or Borden or by developing a marketing niche for specialty cheeses. On-farm commercial cheese production remained, practiced by just a few goat dairies scattered throughout the region that were mostly unnoticed by both consumers and the cheese industry at large. That would soon change.

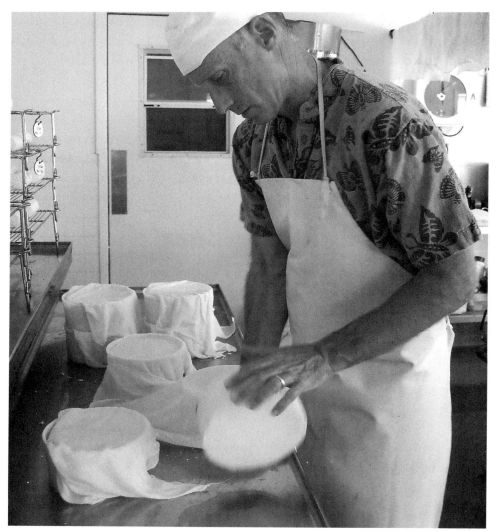
Pierre Kolisch of Juniper Grove Farm in Redmond, Oregon. Photo by the author.

Chapter 6
Back to the Farm:
The Artisan Cheese Renaissance

> So the air is full of crud and the water tastes funny and the
> nine-to-five is a drag. You're tired of the subway, dog crap in
> the streets, bumper-to-bumper traffic and plastic TV dinners.
> Maybe the communes—with all of that fresh air, sunshine,
> love and home baked bread—are onto something.
> *Mother Earth News* #1, January 1970

During the 1960s and '70s communities across the United States began to strain against the restrictive boundaries of racial segregation and Cold War politics. The Vietnam War mobilized students across the nation and college campuses across the Pacific Northwest staged political protests, some of them violent. Alongside the escalating social unrest emerged an intense focus on the natural environment. John Shuttleworth, who along with his wife, Jane, founded the iconic *Mother Earth News* in 1970, dedicated the magazine to two things: "giving people back their lives and stopping the rape of the planet." Rooted in and inspired by the back-to-the-land tradition in the United States stretching back to Thoreau's *Walden*, a new generation sought to re-imagine the planet before society had transformed it into something entirely altered and even dangerous.

The Pacific Northwest's rich and diverse natural landscape sustained its indigenous peoples for centuries and continued to provide a livelihood to later-arriving fur traders and Oregon Trail settlers with its seemingly endless supply of timber, salmon, coal, and precious metals. But by the 1960s and '70s, newly awakened citizens began to face what the intervening decades had wrought. Puget Sound in Seattle and the Willamette River in Portland had become cesspools, loaded down with untreated industrial wastes that had accumulated for decades. More than two dozen dams dissected the Columbia and Snake rivers; while dams brought cheap electric power to the region they also decimated salmon runs. Decades of mining activity had taken its toll in

Idaho, denuding the landscape and leaching arsenic and toxic heavy metals into rivers and streams. Over a century of more or less continuous logging had depleted forests and wildlife habitat all over the region. On many levels, residents of the Pacific Northwest began a process of reframing their regional identity during the latter decades of the twentieth century, and momentum began to shift from the profit-focused exploitation of the area's natural resources toward the determined preservation of what was left.

It was not much of a stretch to connect polluted land and polluted food. In 1974 more than seven hundred people met at the Northwest Alternative Agriculture Conference in Ellensburg, Washington. Woody Dercyx, an instructor at Evergreen State College who helped organize the event, summed up the sentiment behind the gathering: "People have been removed from a healthful relationship with the land and the present methods [of agriculture] destroy the land." Not only had agriculture become harmful, but its products had in turn become tainted and unwholesome, loaded down with pesticides and herbicides. Among those attending the conference were Mark Musick, one of the founders of what became Oregon Tilth, currently one of the leading organic certification agencies in the United States. Also present was Gene Kahn, founder of the original Cascadian Farm (since purchased by General Mills) in Washington's Skagit Valley in 1972. Those at the conference began re-imagining agriculture as a sustainable practice capable of producing healthy food while maintaining a healthy environment. Their collective vision, combined with that of others across the nation, began to build momentum for a new food movement.

The Re-emergence of Farmstead Cheese

The so-called "back to the land" movement that emerged out of the 1960s-'70s counterculture was both an ideology as well as a bona fide phenomenon. Those committed to the philosophy set out on a path of building a life devoted to an idea of self-sufficiency. "The broad aim of the movement," said a reporter for the *New York Times* in a 1975 article, "is to find ways to live simply but well on the land, outside of economic institutions that dominate the United States." For many, living that dream meant acquiring a whole range of practical skills that would have been second nature to previous generations. Volumes of information appeared to show them how, among them the popular *Whole Earth Catalog* series, periodicals such as *Mother Earth News*, *Organic Gardening* (which the forward-thinking J. I. Rodale had been publishing since 1942), and the *Foxfire* book series. The United States Department of Agriculture even got into the act, publishing a guide called *Living on a Few Acres* in 1978. So potent was the phrase "back to the land" in popular culture that real estate companies across the nation used it liberally throughout the decade, running ads exhorting potential buyers to "Move Back to the Land!" and "Back to the Land: Homesteader Special!" And people really did move in large numbers; contemporary demographers noted a remarkable shift in population patterns beginning in the 1970s that they termed a "rural rebound," marking a reversal of population patterns that had been trending toward cities for over a hundred and fifty years. While no good estimate exists of the numbers of people who moved to rural areas during this period for the specific purpose of "getting back to the land," the Census Bureau observed a sizeable net loss of 1.8 million people from urban areas across the United States between the years 1970 and 1974 alone.

Few (if any) of this new generation of farmers started out with the specific goal of making and/or selling cheese, but for many the practice evolved over time. Goats in particular helped usher in a new era of on-farm cheese production. While dairy goats have resided in the Pacific Northwest more or less continuously since the nineteenth century, the growth in popularity of self-sufficiency and sustainable farming practices brought an increase in popular interest in the animals. A 1970 article in *Mother Earth News* titled "Get a Goat!" called goats "an amusing and profitable addition to the homestead." Goats were appealing because they were smaller and easier to care for (especially for agriculture neophytes), required less grazing space, and were far less expensive to keep than cows, but still provided an important source of nutrition. And when this new generation of small farmers needed something to do with their extra goat's milk, some experimented with making cheese.

Rural Polk County, west of Salem, Oregon, has a history of cheesemaking stretching back to the mid-nineteenth century, when Cynthia Ann Applegate made cheese using the milk from the cows her family brought with them over the Oregon Trail. Decades later, during the 1920s, Winnie and Jay Branson produced a goat's milk blue cheese at their Falls City farm. Birgit and Don Blischke joined the area's distinguished cheesemaking lineage when they started Singing Winds Goat Dairy on their thirty-acre farm outside of Dallas, Oregon, in 1978. "We wanted a challenge," remarked Birgit Blischke in a 1984 profile of the farm. "Something weird and difficult. And that's exactly what we got." The former librarian and her husband, who worked in a local mill before quitting to help run the dairy, started out selling raw goat's milk to the local community. They built a business over time and eventually accumulated a sizeable herd of over three hundred goats. A few years later they began producing a goat's milk cheese called Ziegankase (German for "goat cheese"), an aged cheddar-style cheese. Singing Winds developed a significant following over its years of operation and distributed their products as far away as Hawaii and Samoa.

In 1979 Sally and Roger Jackson received a $2,210 grant from the newly minted United States Department of Energy "for the construction and operation of a hand made cheese producing plant designed to operate without electricity or fossil fuels." The Jacksons, who had purchased land in rural Okanogan County, Washington, near the town of Oroville, kept a mixed herd of goats, sheep, and several cows. Sally, the cheesemaker, produced a variety of products including fresh and aged cheeses and ice cream. In search of a customer base, the Jacksons drove their products to Seattle, where they found a few forward-thinking chefs, including Bruce Naftaly of Le Gourmand restaurant in the Ballard neighborhood, who not only appreciated but also purchased her cheeses. By the 1990s, Jackson's cheeses were featured in shops in New York and received multiple raves in the *New York Times*; one reviewer called Jackson's aged sheep's milk cheese "one of the finest cheeses in this country." Jackson later narrowed her focus to the leaf-wrapped cheeses that she became known for: the cow's and sheep's milk versions wrapped in chestnut leaves and the goat's milk cheese in grape leaves.

In rural Graham, Washington, east of Tacoma, David and Jeanne Greatorex started Kapowsin Dairies in 1980. David Greatorex, a native of England, originally met his wife in Tanzania, where she was working for the Peace Corps and he was an engineer for the Tanzanian government. He later got a job with Kaiser Aluminum and the family relocated to the Tacoma area, purchasing a nineteen-acre farm. After spending a few years studying cheesemaking, Greatorex left his job and the family devoted themselves full-time to milking their herd of thirty-six goats and making cheese. Kapowsin's soft goat's milk cheese, Nisqually Capri, and the aged Capricese were sold as far away as New York, Chicago, and San Francisco. "[My cheese] gives people the chance to eat cheese the way it used to be made, but isn't anymore," remarked Greatorex about his products. In 1982, Marian Burros described Nisqually Capri in the *New York Times* as a beige disc-shaped cheese. "With aging, the coat becomes darker and the cheese softens, taking on a sharper, more goatlike flavor."

True to the spirit of the era, the Oregon Dairy Goat Cooperative combined the idea of making cheese with a larger social purpose. In 1971, University of Oregon graduate students Bill and Nancy Ulhorn started the organization under the auspices of the VISTA program, a precursor to the AmeriCorps domestic service program. VISTA volunteers placed goats with families in low-income communities and trained them in the ins and outs of dairy goat care. The co-op developed a membership of about one hundred small-scale goat farmers scattered across western Oregon who provided milk that the organization picked up and trucked to its cheesemaking facility in Salem, Oregon. News reports of the period indicate that the operation struggled from the outset and operated at a continuous deficit, however. Manager Miriam McVickar ran into a number of practical difficulties, including educating the member farmers about goat care, dairy sanitation, and the proper handling of milk. In 1972 the co-op received a grant from the Lane County Community Action Program, but the funds weren't enough to sustain the organization and it closed soon after.

While many of the Pacific Northwest's emerging wave of cheesemakers were making cheese from goat's milk, some dairy farmers with cows also saw opportunity in the cheese business. George Train purchased a farm in 1963 in Ferndale, Washington, near the Canadian border, but soon faced the dilemma of many farmers: how to sustain it. Train ended up taking a job at a local aluminum refinery to support his family and, for a time, put the farm up for sale. He eventually acquired a few cows and began selling milk to a local processor. During the mid-'70s he developed a raw milk business; though raw milk was popular, demand waxed and waned and Train eventually settled on the idea of making cheese as a more sustainable business model. "The bottom line was, we started making cheese out of necessity," recalls Train's daughter Joyce Snook. Train developed a line of cheeses that didn't require expensive equipment, including Dutch-style gouda and a cheese he called Farmstead (which the farm still produces), developed in collaboration with Margaret Morris of Glengarry Cheese Company. Pleasant Valley Dairy continues to produce cheese today under the stewardship of Train's children and grandchildren.

The modern wave of farmstead cheesemakers in the Pacific Northwest was part of a quiet movement spreading across the United States. A growing cadre of small farmers-turned-cheesemakers, many making cheese from goat's milk, slowly began to make inroads in the well-established commodity cheese production industry. Among other things, they returned to practices long since tossed aside by industrial producers, making cheese in small batches from the milk of animals that many raised on their own farms. They made new and delicious styles of cheese, giving consumers choices beyond cheddar and mozzarella, the most-produced cheeses in the industrial sector. And as they grew and prospered, these small cheesemakers helped nurture a market for locally produced cheese and farm products of all kinds. They broke down barriers with retailers and distributors, who soon saw the benefits of supporting and marketing locally made cheeses that were becoming popular with consumers. Those who drew national attention, such as Washington cheesemakers Sally Jackson and David Greatorex of Kapowsin Dairies, furthered their own reputation as well as boosting the profile and credibility of the region's producers generally. Their efforts went a long way toward developing a receptive environment that made future success for others possible.

An Emerging Food Revolution

Those who live in the Pacific Northwest of the twenty-first century might have a hard time imagining a world without the many local coffee roasters, craft brewers, or artisan cheesemakers they've become accustomed to. But just a few decades ago, the region's food landscape looked quite different. When Elaine Tanzer started Elephant's Delicatessen, one of the first "gourmet" shops in Portland, Oregon, in 1979, she had a hard time finding products to stock her new store. She was forced to source cheeses for her shop on trips to New York City and incredibly (at least by current hard-core coffee culture standards) had considerable difficulty locating an espresso machine. Ralph Bolson, owner of Pike Place Cheese, a small cheese shop that operated in the Pike Place Market in Seattle during the 1970s and '80s, was thrilled when he discovered local cheesemaker Kapowsin Dairies in his own back yard. "I've been waiting eight years to find a good goat milk cheese made locally," he remarked in a 1980 *Seattle Times* article. Hubert Loevenbruck, owner of Portland-based Eurobest foods, a wholesale distributor, once said that he decided to start his company in the Pacific Northwest because he saw opportunity in the heretofore cheese-starved region. "Most of the French cheese received here [in the Northwest] was overflow of that brought to New York, good or bad . . . I knew that people here would appreciate better than that."

Consumer tastes were evolving. A new era of travel was opening up the world and the number of U.S. airline travelers rose 377 percent between 1960 and 1980. Europe became easily accessible to a growing middle class of Americans who brought back an appreciation for the foods and wines they enjoyed there and a desire to purchase them at home. It's no accident that the Pacific Northwest wine industry also emerged during this period, as consumer palates began to turn away from then popular fruity and fortified wines toward the drier fine wines of Europe. In 1966, David Lett planted pinot noir grapes near Dundee, Oregon, a moment that launched the state's present-day international reputation for pinot noir wines. A group of professors from the University of Washington started one of that state's first modern-era wineries, Associated Vintners, in the late 1950s and by 1961 Washington was home to eight wineries. A new era in food production and consumption was on the horizon.

Given the growing popularity of all things European, it's no surprise that many of the Pacific Northwest's early cheesemakers brought European names and influences to their cheeses. Pat McCoy, then owner of Nehalem

Bay Winery, started the Blue Heron French Cheese Co. with Denny Pastega in 1980 in Tillamook, Oregon. Though neither was a dairy farmer, the operation played to the area's cheesemaking history and was located near the Tillamook Cheese Factory, a well-established and popular tourist stop along Highway 101. McCoy's first employee was cheesemaker Cindy Morrow, a graduate of Oregon State University's Food Science program. Morrow made the company's signature French-style soft-ripened cow's milk cheese, among the earliest commercially produced bloomy-rinded cheeses in the Pacific Northwest. The cheese was good enough to win a second-place ribbon at the second annual competition of the American Cheese Society in 1986. While the Blue Heron French Cheese Co. no longer makes cheese, the signature white barn continues to serve as a popular tourist destination, proof that the marketing instincts of the owners were correct.

Although many artisan cheesemakers of this era learned to make cheese by trial and error, Pierre Kolisch, a former attorney, learned to make cheese by studying cheesemaking in France for two years, attending the National School of Dairy Technology (ENILBIO) in Poligny and later apprenticing alongside Camembert producer Francois Durand. Kolisch then returned to his native Oregon, purchased five acres in the central Oregon town of Redmond, acquired goats, and started Juniper Grove Farm in 1987. A few hours away in Portland, he found a burgeoning market for his French-style goat's milk cheeses among a group of new Portland chefs, also with European roots and influences, looking to exploit the great bounty of locally grown and produced Pacific Northwest foods. "Pierre was the first cheesemaker I found in the state that I thought was making real cheese," says Greg Higgins, who opened Higgins restaurant, one of the city's first locally focused bistros, in 1994. Higgins knew what he was talking about, having made cheese at a small plant in his hometown of Eden, New York, during his high school years and trained as a chef in France before starting his restaurant.

Four dairy farmers started Yakima Valley Cheese Company in Sunnyside, Washington, near Yakima, in 1985. Dave Newhouse, George Halma, Willard Winters, and Louis Jernecke built the business around the production of Dutch-style cheeses, even bringing in cheesemaker Peter Sikkens from the Netherlands to supervise the operation. Their Yakima Gouda became a popular regional specialty. A few years after it opened, a portion of the company was purchased by Netherlands-based Westland; the company may have seen the Yakima outfit as a foothold to expanding its reach in the United States marketplace. After the acquisition the operation expanded its production to include additional Dutch specialties such as edam and havarti and a variety of flavored goudas. Though the company's cheeses won multiple awards at

American Cheese Society competitions and the cheese plant seemed perfectly situated to complement the area's growing wine industry, the company closed during the early 1990s.

One of the most important influences on the development of the regional artisan cheesemaking community was the parallel growth and proliferation of farmers markets. Despite their popularity during the early twentieth century, farmers markets across the nation had fallen into a steep decline. One important reason was simply the decline of small farms as the twentieth century progressed; big agriculture did not need farmers markets to survive. Changes in the world of retail merchandising including the spread of supermarkets also took a toll; consumer shopping patterns also changed significantly with the post-World War II advent of sprawling suburbs that drew consumers away from city centers. Once bustling community gathering places, public farmers markets became a quaint anachronism.

Farmers markets did not disappear entirely (the public market in Salem, Oregon, which started in 1943, is one notable exception), and by the 1970s they were once again on the rise in the Pacific Northwest. In Seattle, several small markets opened the mid-1970s as part of an effort to connect low-income residents with fresh farm-raised foods, and by 1979 the *Seattle Times* was moved enough to report that "Farmer's Markets Are Blossoming," noting venues springing up across the state from Aberdeen to Redmond to Olympia. During the mid-'70s farmers markets also appeared in both Pocatello and Twin Falls, Idaho. In Oregon, farmers markets started in Albany and Newport in 1978 and the trend spread to the Portland suburban cities of Hillsboro, Gresham, and Beaverton, all of which established markets by the mid-1980s. Downtown Portland finally got its own farmers market in 1992, and in Idaho, the Boise Farmers Market (now called the Capitol City Farmer's Market) was up and running by 1994. By 2012, there were 263 farmers markets across the Pacific Northwest, located in the region's largest cities and in small rural communities up and down the Oregon and Washington coastline, and in small communities in the Rocky Mountains of eastern Idaho.

Farmers markets have grown into important marketing venues for cheesemakers and small food producers of all kinds, serving as small business incubators where cheesemakers can introduce themselves in a receptive retail environment. Rod Volbeda saw farmers markets as a relatively low-cost way to get into the cheese business when he started Willamette Valley Cheese Co. in 2002. "When farmers markets started becoming popular I figured I could get started in the cheese business that way—we couldn't afford to get a big loan." Retailers at farmers markets can also assess consumer interest and receive immediate feedback about their products, and the consumer relationships and name recognition developed there can also go a long way toward ensuring a new cheesemaker's long-term success. "Our farmers market booth is also our public face and mobile storefront. It conveys our image beyond our label," says Sarah Marcus of Briar Rose Creamery in Dundee, Oregon. "That helps build up customer loyalty, too. Hopefully, [customers] get a good feeling meeting us, like what they see, and will come back week after week to buy our cheese." Perhaps more importantly, at farmers markets cheesemakers can earn money immediately with no marketer or distributor taking a cut out of their profits, a fact that may be especially critical to a startup business venture.

Another important factor in the growth of the artisan cheese movement has been the spread of retail shops specializing in gourmet products of all kinds, including cheese. Seattle grabbed worldwide attention and dollars in the 1990s with nationally expanding home-grown coffee chain Starbucks, dominant global software company Microsoft, headquartered in suburban Redmond, Washington, and its wildly popular grunge music scene. Economic expansion led to a boom in the city's food and restaurant community and a new generation of shops and restaurants marketed the idea that high-quality foods, both local and international, were not only tasty but important and worth celebrating. Cheese-minded consumers were able to choose from an array of retail venues including Brie & Bordeaux near Green Lake, Fromagerie in Madison Park, and Quality Cheese in the Pike Place Market.

People still talk about the James Cook Cheese Shop, a small store that was open for a few years on Second Avenue in Seattle. James Cook, a Scotland native, had developed a reputation while working as a cheesemonger at DeLaurenti's Italian Market in the Pike Place Market. By all accounts Cook

Cheesemaker Sarah Marcus inside the curing room at Briar Rose Creamery in Dundee, Oregon. Photo courtesy Christine Hyatt.

was a charismatic man whose connections with Neal's Yard Dairy cheese shop in London enabled him to bring cheeses to the Pacific Northwest the likes of which few had ever seen. Cook made a big splash in the city at a time when urban Seattleites were feeling flush and embracing the finer things in life. Theresa Simpson and Dennis Nelson befriended Cook and helped him find storage space for his growing inventory of cheeses, and later provided financial backing for his eponymous cheese shop. "No one was doing what James was doing," says Simpson. "Farmstead, artisan cheeses—all strictly British Isles. He had personal stories about each farm and cheese maker. At events, they would haul out the cutting table and cut wheels of cheese on the spot. People would stand in line for two or three hours just to get a sample. You had to see it to believe it." While Cook proved unable to channel his considerable charisma into ongoing business success, he succeeded in proving that Seattle consumers could be wildly enthusiastic about good cheese, if given the chance.

As the artisan cheesemaking community grew and consumers became increasingly captivated by the wonders of local and farm-produced cheese, the nature of the cheesemaking business in the Pacific Northwest began to change. No longer just the signature hobby of counterculture hippies, cheesemaking, by the advent of the twenty-first century, began to assume the mantle of a respectable business proposition. David Gremmels and Cary Bryant were successful entrepreneurs looking to start a wine bar in Ashland, Oregon, before they purchased Rogue Creamery in Central Point, Oregon,

Beecher's Handmade Cheese of Seattle opened a second urban creamery in 2011, in New York City. Photo courtesy Beecher's Handmade Cheese.

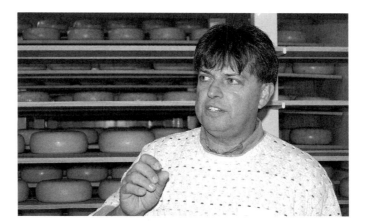

Flavio DeCastilhos, Tumalo Farms Cheese Co., in Bend, Oregon. Photo by the author.

from the Vella family in 2002. They've since brought the once-struggling creamery back into the limelight as one of the Northwest's most well-known and successful cheese producers. Beecher's Handmade Cheese, a unique urban creamery in downtown Seattle's Pike Place Market, opened in 2003. Entrepreneur Kurt Dammeier, owner of several Seattle area food businesses, grew the company over the next several years, producing a number of cheeses including its popular Flagship, modeled after Washington State University's Cougar Gold (WSU also happens to be Dammeier's *alma mater*). In 2011, Beecher's debuted a second location in New York City. Likewise, Flavio DeCastilhos brought his considerable business skills honed in Silicon Valley to the world of artisan cheesemaking when he started Tumalo Farms in Bend, Oregon, in 2005. DeCastilhos has gone on to develop the largest goat dairy in Oregon and specializes in aged gouda-style goat's milk cheeses like the popular Classico, found in retail shops across the country. The success of these operations reflects the business acumen of their owners, and the consistency and quality of their cheeses has furthered the reputation of the Pacific Northwest cheesemaking community both locally and nationally.

The rapidly increasing supply and diversity of artisan cheeses has met a growing demand. During the decade between 2000 and 2010, another wave of cheese-focused retailers appeared on the scene. Theresa Simpson and Dennis Nelson opened their own shop, The Cheese Cellar, in Seattle in 2004. In Portland, three cheese-focused shops opened in 2005: Steve's Cheese, owned by Steve Jones (which has since evolved into a cheese shop/restaurant venue called Cheese Bar), Foster & Dobbs, started by Luan Schooler and Tim Wilson, and Curds & Whey, which has since closed. These retailers have been instrumental in nurturing the continued growth and prosperity of the artisan cheese movement, developing relationships with cheesemakers and actively promoting the region's cheeses to consumers hungry for local products of all kinds. It helps that today's cheesemongers have a plethora of high-quality local cheeses from which to choose. "When I first started Steve's Cheese, if we had a dozen local cheeses in our case we were doing really good. Now from the greater region we might have thirty or more available at any given time," says Steve Jones. Cheesemongers have played an important role in educating consumers, offering workshops and classes covering local, national, and international cheesemakers, cheese styles, and regions as well as food and beverage pairings. Cheese shops are opening in smaller cities all over the Pacific Northwest, including Cheese Louise in Richland, Washington; Saunders Cheese Shop in Spokane, Washington; and Abbie and Oliver's in McMinnville, Oregon. Cheese-focused retailers have emerged as a vital link in the marketing chain between cheesemakers and consumer, and their proliferation across the region signals a healthy cheese economy.

Idaho Goes Its Own Way

During the nineteenth and early twentieth century, the state of Idaho lagged considerably behind Oregon and Washington, at least when viewed through the lens of dairy production. Oregon set the regional standard for both milk production and butter and cheese manufacturing well into the twentieth century. That state of affairs has since changed dramatically; Idaho is now the third-largest dairy state in the United States, trailing only California and Wisconsin, perennial competitors for the first place title. In 2010, dairy plants in the state of Idaho produced 849,568,000 pounds of cheese—that's nearly *one billion* pounds of cheese.

Idaho's dairy industry has always depended to a large extent on the kindness of strangers. Throughout the state's history, outsiders such as J. L. Kraft from Illinois, Wisconsin's Laabs family, the Nelson-Ricks Co. and Gossners from Utah, and Chicago-based Swift and Co., among others, have all exploited Idaho's dairy bounty. Homegrown potato magnate J. R. Simplot made a play at the regional dairy business in the 1990s when his J. R. Simplot Company purchased a number of mid-sized cheese operations, but sold its dairy stake within a few years. Fueled by an influx of dairy farmers from the more heavily regulated state of California, the state's milk production has surged in recent years. The growth has, in turn, drawn the attention of international dairy companies such as France-based Groupe Lactalis and Ireland-based Glanbia. As of 2012 most of the major players in Idaho dairy circles are based outside the state and the largest are based outside the country. Idaho is poised at the forefront of a developing trend—the globalization of the dairy industry.

Here is a brief portrait of Idaho's commodity dairy products industry as of 2012:

Brewster Cheese Co. In 2006, Ohio-based Brewster Cheese Co. purchased a Kraft-owned production facility in Rupert, Idaho. The plant, now known as Brewster West, is the exclusive producer of skim barreled cheese (an industry term for bulk cheese packaged in barrels for further processing) for the Kellogg Company, which uses the cheese to make Cheez-It crackers. Brewster operates two additional plants in Brewster, Ohio, and Stockton, Illinois.

Davisco. The Jerome Cheese Company in Jerome, Idaho, is operated by Minnesota-based Davisco, a large international dairy company that owns two other dairy processing facilities in Minnesota and South Dakota. Davisco is one of the primary cheese suppliers for Kraft Foods, and produces a combined three hundred and seventy million pounds of cheese per year.

Glanbia Foods. Irish dairy firm Avonmore purchased Idaho-based Ward's Cheese Co. in 1990. Avonmore later evolved into Glanbia plc, a dairy company with operations that span the globe from the United States to Europe, China, and Nigeria. Glanbia operates four dairy products plants in Idaho. Two cheese plants, located in Twin Falls and Gooding, produce a variety of cheeses including cheddar, jack, and colby and two additional plants process whey into protein supplements and food additives. Glanbia's Idaho operations produce over four hundred million pounds of cheese and one hundred and ten million pounds of whey-based dairy ingredients annually. Glanbia also owns Southwest Cheese Co. in Clovis, New Mexico, in partnership with the Greater Southwest Agency.

Gossner Foods. Swiss-born Ed Gossner made cheese for the Cache Valley Dairymen's Association in Utah before forming his own company in the 1960s. Today Utah-based Gossner foods operates three large cheese plants in the western United States: one in Logan, Utah, one in El Centro, California, and a third in Heyburn, Idaho, just east of Twin Falls. The Heyburn plant opened in 2005 in a converted potato processing plant and currently produces swiss cheese.

Litehouse Foods. Litehouse began as a restaurant run by Ed and Lorena Hawkins in the tiny resort town of Hope, Idaho, on the shores of Lake Pend Oreille. Today Litehouse Foods is an employee-owned operation based in Sandpoint, Idaho, with sales reportedly in the neighborhood of $100 million annually. The company's Sandpoint plant produces blue cheese under the supervision of longtime cheesemaking master Ralph Stuart, who once worked in the blue cheese caves at Faribault, Minnesota. Another plant in nearby Kootenai produces buttermilk for the company's salad dressings. Two additional plants in Lowell, Michigan, and Hurricane, Utah, produce a range of products including salad dressings, dips, marinades, and cider.

Nelson-Ricks. Founded in 1907, this Utah-based company has had a presence in Idaho since the early twentieth century and operated as many as eleven cheese plants in the state during the 1920s. In late 2012, the company announced that it was closing its last Idaho factory in Rexburg, where it had been producing a variety of cheeses including cheddar, monterey jack, and colby sold under the Banquet and Grand Teton brands. The company continues to operate a plant in Salt Lake City, Utah.

Sartori. Since its founding in 1939, Wisconsin-based Sartori has developed a lucrative niche producing specialty Italian and Mexican-style cheeses. In 2006 it purchased the Blackfoot Cheese Co. of Blackfoot, Idaho, where it produces Italian-style hard aged cheeses such as parmesan and romano. Privately held Sartori also operates three other cheese manufacturing plants, all in Wisconsin, as well as another facility in Colorado that manufactures dips and spreads.

Sorrento Lactalis. Sorrento Lactalis is a subsidiary of France-based Groupe Lactalis. Its Nampa plant, on the site of the former Swiss Village Cheese Co., currently produces primarily mozzarella and string cheeses. In 2012, Lactalis announced a planned $40 million expansion of the Nampa facility that would allow the company to dramatically expand its production of fresh mozzarella from seven million to forty million pounds annually. With the addition, the plant would become the largest Lactalis-owned plant in the world.

Idaho's dairy products industry is not solely cheese focused. Wisconsin-based Marathon Cheese Company operates a 212,000-square-foot cheese packaging facility in Mountain Home, Idaho, southeast of Boise, one of three

Sorrento Cheese Factory, Nampa, Idaho. Photo courtesy Gerry Slabaugh.

so-called "cut and wrap" operations it owns across the nation. The Commercial Creamery Co. captures a bit of a different dairy market: its Jerome, Idaho, plant processes milk into a variety of flavorings, powders, and seasonings used on chips and in salad dressings, sauces, and soup bases. Seattle-based Darigold operates three plants in Idaho out of a total of thirteen dairy products plants in four western states. In addition, two Idaho-based companies, Idaho Milk Products in Jerome and High Desert Milk in Burley, operate processing plants that turn milk into a variety of powdered milk and milk protein products for the food industry. In December of 2012 the Chobani Company, makers of a popular brand of Greek-style yogurt, opened a new 940,000-square-foot yogurt manufacturing plant in Twin Falls, Idaho, the company's second plant (the other is located in New York).

While industrial-scale cheese production in the Pacific Northwest is centered in Idaho, it's worth mentioning that Washington and Oregon are each home to one remaining large-scale cheese production company. Darigold has operated a cheese plant in Sunnyside, Washington, near Yakima, since 1996 that produces about one hundred seventy-five million pounds of cheddar cheese per year. In 2001, the Tillamook County Creamery Association, based in the coastal town of Tillamook, Oregon, opened a second production facility in Boardman, a small town located over two hundred miles away on the Columbia River in eastern Oregon. The area was attractive to the company in part because it is home to Threemile Canyon Farms, a dairy operation home to more than forty thousand cows. In 2007, Tillamook expanded its Boardman cheesemaking facilities, doubling their size. Though the plant in Tillamook still makes cheese, today the majority of Tillamook's cheese is produced at the Boardman plant. In early 2012 the Tillamook County Creamery Association announced that it was phasing out packaging operations at its home facility on the Oregon coast and that those operations would be outsourced to companies in Idaho and Utah. The company's move makes it clear that it is an open question how long industrial-level cheese production can remain financially viable on the Oregon coast.

Despite the overwhelming dominance of industrial cheese production in Idaho, a small but growing group of artisan cheesemakers has emerged in the state. As was the case in Oregon and Washington, Idaho's earliest small-scale farmstead cheesemakers were goat farmers. Pioneers Chuck and Karen Evans started Rollingstone Chevre in Parma, Idaho, in 1988, among the first farmstead cheesemakers the state had seen since Oregon Trail days. Parma, Idaho, is located on the western edge of the state near the Oregon border and is the former site of the Hudson's Bay Company's Fort Boise. The Evans family began raising goats in Rollingstone, Minnesota, after finding that one of their daughters was allergic to cow's milk. They later moved to the Parma farm that had been in Chuck's family for decades, and took up cheesemaking in earnest. Karen's expertly crafted cheeses were on the leading edge of the growing artisan cheese trend and Rollingstone soon developed a national following. Several new goat's milk cheesemakers have appeared on the scene in recent years, including Wheyward Goat Cheese Co. in Priest River in the northern panhandle and Green Goat Dairy in Shoshone in south central Idaho near Twin Falls.

Currently just a few companies in the state produce small-batch cow's milk cheeses. Gooding, Idaho, home to cheese giant Glanbia Foods, is also home to the far smaller Ballard Family Dairy and Cheese Company. Steve and Stacie Ballard relocated to Steve's native Idaho from California during the 1990s and started a dairy farm. Fearful that they would not be able to survive as a small dairy, the Ballards constructed a cheesemaking plant and started making cheese in 2004. Today they keep a herd of about eighty Jersey cows, selling some of their milk to a local processor and using the rest to produce a line of artisan cheeses including cheddar, feta, and their popular Idaho Golden Greek Halloumi grilling cheese. Newcomer Teton Valley Creamery, located in Driggs, Idaho, near the Wyoming border, started in 2010, resurrecting a dormant cheese industry in the Rockies of eastern Idaho where swiss cheese once reigned supreme. The operation uses milk from a local dairy herd to produce the alpine-style Yellowstone, mellow Haystack, and Sapphire Blue, a blue cheese produced in small 1.5-pound wheels.

One of the more recent trends in Idaho cheese circles is the blossoming of a sheep's milk cheese industry. The state has a long history of sheep ranching stretching back to the nineteenth century. Many immigrants from the Basque region of Spain migrated to the United States and worked in Idaho's sheep industry as shepherds and ranchers, raising sheep for meat and wool. Today Idaho is home to the largest concentration of people of Basque ancestry outside of Spain. Sheep ranching continued to be a profitable endeavor throughout the twentieth century, in fact, so plentiful were sheep in the state that it is often said that the sheep population exceeded the human population until the 1970s. During the twenty-first century two sheep's milk cheese producers have joined the state's artisan cheesemaking scene: Lark's Meadow Creamery in Rexburg in the eastern part of the state and Laura Sluder's Blue Sage Farm in Shoshone, Idaho. Lark's Meadow has had cheesemongers across the country taking notice, and cheesemaker Kendall Russell's aged Dulcinea cheese won several awards at the American Cheese Society competition in 2011.

One reason that Idaho is home to relatively few small artisan cheesemakers compared to Oregon or Washington is that the state has few big population centers and is far away from consumers in larger cities such as Seattle, Portland, and/or San Francisco, a fact that complicates the task of marketing farm products such as cheese. Distant markets have always been a challenge for the state's dairy industry; even during the 1930s, more than three quarters of cheese made in Idaho was shipped outside the state, most to California. Boise, Idaho's largest city, is relatively small—as of 2010, Boise had 205,671 residents and just over six hundred thousand people lived in the surrounding metropolitan area. Compare that to Seattle, the Pacific Northwest's largest city, with 3.4 million people in the metropolitan area, and Portland with 2.2 million in the surrounding region, and the disparity is clear. "Idaho is still very rural, except for Boise," says Stacie Ballard of Ballard Family Dairy and Cheese Company. "Our main markets for selling cheese are Boise, Sun Valley, and Coeur d'Alene in Idaho and to the east in Jackson, Wyoming." Susanne Wimberley of Wheyward Goat Cheese Co. near Priest River, Idaho, in the northern part of the state, markets her cheeses in Sandpoint, one of Idaho's ski resort towns. She also takes advantage of the shops and farmers market about an hour to the west in Spokane, Washington (population 472,147 in 2010), a city that offers a much broader consumer base for her products. The relative dearth of consumers in the state will act as a perpetual damper on the growth of Idaho's small-scale artisan cheesemaking community.

Current Trends

Jeffrey Roberts estimates that in 1990 there were about seventy-five artisan cheesemakers in the United States. By 2006, the year that he published the *Atlas of American Artisan Cheese*, he had discovered over four hundred. As of 2012, Roberts believes there are approximately eight hundred and fifteen artisan cheesemakers operating across the nation. The growth of the Pacific Northwest's artisan cheesemaking community reflects this national growth curve. In 2002, Washington had a small community of just fourteen artisan cheesemakers, Oregon had five, and Idaho two. As of 2012 Washington was home to the largest number of artisan cheesemakers in the Pacific Northwest, fifty-two; Oregon boasted twenty-five and Idaho ten. The rapid expansion of artisan cheesemaking over the past several decades has been truly remarkable.

The artisan cheesemaking community in the Pacific Northwest has grown to become an eclectic, diverse group. More than half of the region's small cheesemakers are making cheese, either entirely or in part, with goat's milk. At least ten are making cheese either entirely or in part from sheep's milk; among these are Ancient Heritage Dairy in Madras, Oregon, and Glendale Shepherd on Whidbey Island in Washington, as well as two in Idaho. Several cheesemakers, including Stephen Hueffed and Amy Turnbull of Willapa Hills

Farmstead Cheeses in Doty, Washington, and Rod Volbeda of Willamette Valley Cheese Company in Oregon, are combining sheep's milk and cow's milk to make delicious mixed-milk cheeses. While many distribute their products nationally, others such as Mama Terra Microcreamery in southern Oregon or Mystery Bay Farm in tiny Nordland, Washington, on the Olympic Peninsula, make cheese that is likely to be found only at a local co-op or farmers market in the vicinity of the farm.

Since artisan and farmstead cheese production started building significant momentum all over the United States in the 1970s, cultural awareness and emphasis on the importance of local, sustainably grown foods has expanded considerably. The 2005 book *The 100 Mile Diet*, by Alisa Smith and J. B. MacKinnon, two British Columbia authors who for one year ate only food caught or grown within a hundred miles of their home, was instrumental in galvanizing what has come to be called the locavore movement. Michael Pollan's 2006 book, *The Omnivore's Dilemma*, reached a broad audience of American consumers seeking more healthful food choices. The local food movement has fueled the public's voracious appetite for locally produced products of all kinds, including artisan cheese—in short, the market continues to grow.

The Pacific Northwest's agricultural universities continue to provide guidance and support to the growing artisan cheese industry as they have since their inception over a century ago. Washington State University revived its artisan cheese short course series in 1985 under former creamery manager Marc Bates. And of course the university still produces and sells Cougar Gold as well as several other styles of cheese, all packaged in the creamery's signature metal cans. In 2010 the Oregon State University Food Science Department debuted a state-of-the-art dairy center thanks to a generous donation from Paul Arbuthnot, longtime president of Sunshine Dairy Foods in Portland, and his wife, Sandra. Among other things, the endowment enabled OSU to revive the university creamery that closed in 1969, and in 2012, OSU debuted Beaver Classic, an aged alpine-style cheese produced by students in the university's food science program. Under the leadership of Professor Lisbeth Goddik, OSU has initiated an innovative cheesemaker business incubator program in which prospective cheesemakers admitted to the program can make cheese at OSU's facility, perfecting techniques and receiving practical guidance before setting out on their own.

In 2006, Oregon's artisan cheesemakers founded the non-profit Oregon Cheese Guild. In addition to sponsoring several cheese festivals and an annual education program, the guild has grown into a vocal marketing organization advocating for the state's cheesemakers, touting the benefits of Oregon-made

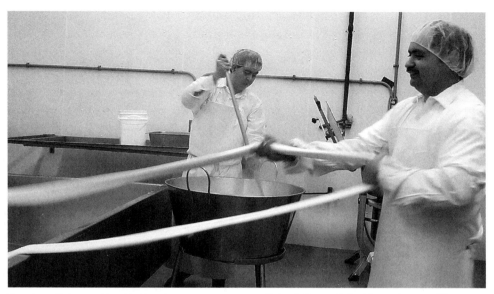

Francisco Ochoa making his signature queso oaxaca at Ochoa's Queseria in Albany, Oregon. Photo courtesy Christine Hyatt.

cheese at trade shows and food events. Working together, Oregon's cheese-makers have made a strong impression on retailers and consumers, reaching a broader audience than most members would have been able to individually. The guild has also benefited tremendously from a unique partnership with the Dairy Farmers of Oregon, an organization that administers dairy-centric marketing and educational programs for the Oregon Dairy Products Commission. While, in most states, milk marketing organizations focus exclusively on promoting cow's milk dairy products (for which each state's cow dairy farmers are charged a premium), the Dairy Farmers of Oregon under Executive Director Pete Kent has recognized that the state's artisan cheesemaking industry contributes positively to the state's dairy reputation, whether the milk comes from goats, sheep, or cows.

A small group of artisan producers of Mexican-style cheeses has emerged in the Pacific Northwest serving the region's growing Hispanic community. As of 2012 four regional cheesemakers specialize in producing these cheeses. In Albany, Oregon, Francisco Ochoa of Ochoa's Queseria makes a variety of Don Froylan brand cheeses including queso fresco and queso oaxaca. Ochoa has been making cheese since 2003, starting out in Eugene before moving to its current location in Albany. Ochoa's queso oaxaca took a third place ribbon in its category in 2012 at the American Cheese Society competition. Two Washington producers currently make Mexican-style cheeses: El Michoacano in Marysville, Washington, north of Seattle, owned by Maria and Gabriel Montes, and Queseria Bendita in the Yakima area, owned by Benedita Aguilar. Their cheeses are distributed primarily through Mexican markets and restaurants in Oregon, Washington, and Idaho. In general, the Pacific Northwest's Hispanic cheesemakers have grown to serve the needs of their communities, much like the Italian cheesemakers of decades past, but to date they have been mostly overlooked by many locavores and artisan cheese consumers. But if the growth pattern for Mexican cheeses follows that of Italian cheeses in past decades, that could change in the future.

In many ways, the artisan cheesemaking industry in the Pacific Northwest is coming of age. Oregon's Rogue Creamery's well-known Rogue River Blue cheese has twice taken the top prize, Best in Show, at the American Cheese Society competition, in 2009 and 2011. Seattle's Beecher's Handmade Cheese took the top honor in 2012 with its Flagsheep, an aged cheese made with a blend of cow's and sheep's milk, marking the third time in four years that a cheesemaker from the Pacific Northwest took the organization's top honor. Numerous other area cheesemakers have accumulated awards for their outstanding cheeses, including Pat Morford of Rivers Edge Chèvre in Logsden, Oregon, and Mt. Townsend Creamery in Port Townsend, Washington. Northwest cheesemakers are using the milk of cows, goats, and sheep to make an appealing, diverse variety of cheeses from delicate bloomy-rinded goat's milk specialties to leaf-wrapped beauties to aged Dutch-style goudas and rich, nutty alpine cheeses. Artisan-made cheeses from the Pacific Northwest are sold all over the United States and are mainstays at nationally known cheese-centric retailers like Murray's in New York City and in nationwide grocery chains such as Whole Foods. So popular is the artisan cheese movement in the Pacific Northwest that a number of cheese-focused festivals have appeared in the region including the Oregon Cheese Festival in Central Point, Oregon, the Wedge Festival in Portland, and the Washington Artisan Cheesemakers Festival in Seattle, which held its inaugural celebration in 2012. Each of these events helps spread awareness of the region's cheesemakers to an ever-broader community of consumers.

The Future of Artisan Cheese in the Pacific Northwest

In 2010 total United States cheese production exceeded ten billion pounds. A substantial chunk of this production is mozzarella cheese, which in 2002 overtook cheddar as both the most produced and consumed style of cheese in the United States. While the USDA offers volumes of statistics on myriad aspects of the commodity cheese industry, there is as of yet no definitive measure of national artisan cheese production or consumption in the United States. Author Jeffrey Roberts estimates current total domestic production of artisan cheese at around fifteen million pounds per year, a tiny fraction of

Toledo Cheese Days, Toledo, Washington

Tiny Toledo is located in the southwest part of Washington along the Cowlitz River. Dairying and cheesemaking have been a part of the region's history for nearly two hundred years; the town is located at the site of the former Cowlitz Farm operated by the Hudson's Bay Company's affiliated agricultural operation, the Puget Sound Agricultural Company. When the Hudson's Bay Company left the area after the 1846 United States boundary was established, eager settlers carved out homesteads in the area.

In 1918 area dairy farmers organized the Cowlitz Valley Cheese Association and built a cheese factory in the center of the small town of Toledo. The factory turned out to be so successful that just a few months after opening the association ordered a larger cheese vat that would allow them to expand their production to twelve hundred pounds of cheese per day. On June 15, 1919, the members held a celebration to mark the factory's opening, and town merchants organized a program of events. The association distributed the first checks to its members and also provided a free lunch that reportedly consisted of cheese, crackers and coffee. The third annual celebration, held in 1921, had grown enough to be termed a "cheese carnival" by the local press and featured an automobile parade and a baseball game.

Arthur Karlen, owner of Valley Creamery of Oakville, Washington, took over the Toledo factory in 1924. Karlen's dairy empire eventually grew to include a plant at Clatskanie, Oregon, as well as one on Puget Island in the Columbia River near Astoria. After the factory in Toledo burned down in 1928, Karlen opened a new one the next year and the Cheese Days celebration continued annually through the 1930s. By 1940 only Karlen's Oakville and Toledo plants were still in operation and the Toledo plant had ceased making cheese, producing powdered milk and butter instead. The Toledo plant finally closed for good in 1945.

The town of Toledo nevertheless continued to hold the Cheese Days festival off and on in the ensuing decades. In 1975, the Toledo Jaycees ushered in a full-fledged revival of the tradition of an annual civic celebration with a cheese theme. Today, the Cheese Days celebration is held every year on the second weekend in July. In its current incarnation, the four-day event includes a parade, a car show, and a frog-jumping contest—and free cheese sandwiches for all comers.

the nation's overall output. Around two million pounds annually or so, says Roberts, comes from artisan cheesemakers in the Pacific Northwest.

Despite the relatively small scale, the growing community of artisan cheesemakers has made waves within the industry, challenging standards and practices developed in the decades of industrial expansion. During the twentieth century, large manufacturers grew by streamlining production techniques toward a goal of simpler and more efficient mass production. But there were consequences to that approach to making cheese; as scientist and historian Paul Kindstedt has pointed out, technologies developed to standardize the many variables of milk and cheese production has had the ancillary effect of "[making] it more difficult to achieve the full range of flavor and character evident in . . . cheeses made by traditional methods." Among the practices being revived by small-scale farmstead and artisan cheesemakers in the last several decades is the production of raw milk cheeses, which they argue are more complex, flavorful, and reflective of the geography of their region than their mass-produced, standardized counterparts made from pasteurized milk. But this renewed focus on complexity and quality in the artisan cheesemaking community has not been without controversy. In recent years the FDA has indicated that it is in the process of reviewing its rules regarding cheese production, holding out the possibility that, among other things, it may extend the sixty-day aging rules currently in effect for various types of cheeses made from raw milk. This has caused many to worry over the future of artisan cheesemaking in the United States.

Raw milk cheese has developed into a particularly contentious issue in the Pacific Northwest. In late 2010, Washington cheesemakers Estrella Family Creamery and Sally Jackson fell under the scrutiny of federal regulators. In the Estrella case, the FDA inspection uncovered a number of regulatory violations at the family's cheesemaking facility, including the existence of listeria (a genus of bacteria that can cause serious illness and even death) in the creamery's facilities and in its aging cheeses; though no one was reported to have become sick the FDA seized the creamery's products, effectively closing the operation. In the Sally Jackson case, several people became ill and the cause was traced to *E. coli* bacteria in Jackson's cheese. Both Estrella Family Creamery and Sally Jackson have since closed their cheese businesses. The close timing of the Jackson and Estrella incidents along with another case of *E. coli* contamination soon after, traced to California's Bravo Farms, sparked a national debate about raw milk cheese and its potential impact on public health that has yet to be resolved.

The state of Idaho has taken a unique approach in regulating raw milk and raw milk cheese. Under what is termed the small herd exemption to its dairy regulations, small farmers with just a few animals may legally produce and sell raw milk and raw milk cheese commercially without setting up an elaborate and expensive Grade A dairy and/or cheese plant. While a number of other states allow for a similar small farm exemption, in Oregon those who avail themselves of this option are not allowed to market their products commercially. Washington requires raw milk producers to be licensed and inspected like traditional dairies. Under Idaho's rules, state authorities periodically test milk and dairy products made by those falling under the state's small herd exemption, but these producers are not subject to the more comprehensive facility testing and inspections that larger operations receive. After the law was passed in 2010, dozens of farms across Idaho were qualified by the state to sell raw milk under the liberal exemption. Several, including Little Bear Dairy in Troy, near Moscow, and Paradise Springs Farm in Victor, near the Wyoming border, are also selling a variety of raw milk cheeses. Little Bear Dairy sells fresh cheeses aged less than sixty days, a product expressly prohibited under FDA rules. While Idaho's regulations could spur the development of a highly localized dairy and cheese economy, it's not clear whether that model will be economically sustainable for farmers in the long run.

Among the other significant issues the artisan cheese industry faces as it moves into the future is the ever-increasing cost of getting into the cheese business. According to Gianaclis Caldwell, cheesemaker at Pholia Farm in southern Oregon and author of the influential *Farmstead Creamery Advisor*, the financial commitment necessary to construct an artisan cheese plant ranges from $100,000 for a simple plant that will likely require later upgrades to $300,000 or more for a larger, more functional facility—and even the higher estimate assumes some sweat equity to keep construction costs down. Neither figure includes the cost of land, animals, and/or purchasing milk, significant components of any artisan cheese business. These considerable barriers to entry are certain to slow the growth of artisan cheesemaking in the future, both regionally and nationally.

For those in business already, the challenge going forward will be to stay in business. Farmstead cheesemakers shoulder the added responsibilities of caring for animals and all that entails: milking (typically twice daily) and maintaining the health and welfare of their animals are significant expenses in an already expensive business. Add to the list of challenges the ever-increasing cost of feeding animals. "Alfalfa has become scarce the past several years because farmers are planting more wheat and biodiesel crops over forage. My feed costs have more than doubled in the past couple of years," says Carine Goldin of Goldin Artisan Goat Cheese in Molalla, Oregon.

In the Pacific Northwest, many cheesemakers also travel long distances from their rural farms to lucrative farmers markets in urban areas.

Another ongoing challenge for the region's cheesemakers will be marketing their wares in an increasingly crowded marketplace. Both veterans and newcomers feel increasing pressure to stand out from the ever-expanding number of artisan cheesesmakers locally and nationwide, all of which must compete for consumer attention and precious retail shelf space. Some farmers markets in Oregon and Washington have reached a saturation point and are no longer accepting artisan cheese producers because they already have so many among their vendor mix. Regionwide, many cheesemakers are turning to social media, including blogs, Facebook, and Twitter, to raise awareness of and generate enthusiasm about their farms and products—potent marketing tools not available to previous generations. In early 2012, Vicky Brown of Little Brown Farm on Whidbey Island, Washington, used the social media funding platform Kickstarter to raise over twenty thousand dollars to develop an aging cave and space to conduct cheesemaking classes, distinguishing the farm as among the first artisan cheesemakers to successfully raise money in this manner.

Innovation will be key to long-term survival for the region's growing community of cheesemakers. Rick and Lora Lea Misterly of Quillisascut Farm in Rice, Washington, started making goat's milk cheese on their thirty-six-acre farm in 1987 after attending the cheesemaking short course offered by Washington State University. In 2002 the Misterleys developed the Quillisascut School of the Domestic Arts, which has since grown into a highly respected farm school that draws chefs and food enthusiasts from across the country interested in immersing themselves in a farm and food experience. A 2008 book about the farm, *Chefs on the Farm: Lessons and Recipes from the Quillisascut Farm School of the Domestic Arts*, authored by Lora Lea Misterly with Chef Karen Jurgensen, chronicled a year at the farm school along with season-specific recipes. Quillisascut also operates a Community Supported Agriculture program in which participants pay up front and receive in return a dedicated amount of farm-produced cheese throughout the year. Kurt Timmermeister of Kurtwood Farms in Washington, who had become known for his delicious soft-ripened Dinah's Cheese, published *Growing a Farmer: How I Learned to Live Off the Land* in 2011. The book, which has garnered Timmermeister national attention and a spot on the *Martha Stewart Show*, chronicles his path to discovering and nurturing his inner farmer and cheesemaker. Across the Pacific Northwest, an increasing number of artisan cheesemakers are attracting customers by offering on-farm stores, farm stays, seasonal open houses, and cheesemaking classes. Others have distinguished themselves by diversifying

Lora Lea and Rick Misterly started Quillisascut Cheese Co. in Rice, Washington, in 1987. Photo by the author.

their product line. In Ferndale, Washington, Appel Farms has found success producing two unique specialty cheeses: paneer, an Indian-style cheese, and quark, a fresh cheese resembling yogurt. David Peterson of Oak Leaf Creamery in Grants Pass, Oregon, makes a bloomy-rinded goat's milk cheese using thistle rennet, reviving a centuries-old style of cheesemaking indigenous to Spain and Portugal. Cheesemakers in the Pacific Northwest are also producing a number of other unique dairy specialties, among them Briar Rose Creamery's chocolate goat cheese truffles, Portland Creamery's cajeta (goat's milk caramel) and Gothberg Farms' chevre cheesecake.

Even as the Pacific Northwest's artisan cheese movement has developed over the past several decades, it must be said that the region's reputation as a cheesemaking center does not yet approach that of more well-established dairy and cheesemaking states such as Wisconsin or Vermont. Wisconsin, for example, has been known nationally as a dairy and cheese-producing state for nearly a century—the slogan "America's Dairyland" graces its auto license plates—and the state is currently home to more than seventy artisan cheesemakers. While there's no question about the connection between Wisconsin and cheese, an average consumer is likely more familiar with the Pacific Northwest's coffee roasters and microbrewers than its cheesemakers. This is at least in part because the reputation of the Pacific Northwest's diverse group of cheese producers, while generally positive, has tended to remain confined within the area. Many of the region's artisan cheesemakers are small and their products are only distributed locally or regionally. For these cheesemakers, the prospect of making a splash in the national marketplace may be secondary to the more immediate day-to-day proposition of making a living and/or sustaining their individual farms and livelihoods. But as Laura Werlin, cheese expert and author of five books including *Laura Werlin's Cheese Essentials,* points out, "the very thing that is creating the Pacific Northwest's reputation as a place where world-class cheese is made is also the main thing that will keep its cheeses below most people's radars: scale." While larger regional producers such as Rogue Creamery in Oregon and Beecher's Handmade Cheese in Washington have grown to serve as standard bearers of the region's reputation, that will only carry so far in the broader marketplace without a larger and more diverse group of producers to sustain it.

The key to a prosperous future for the Pacific Northwest's artisan cheesemaking community could come in working together. Some of the Pacific Northwest's newest generation of cheesemakers have been busy in the past few years creating collaborative cheesemaking models. In south-central Idaho, Jillian Greenawalt of Green Goat Dairy forged a partnership with Laura Sluder of Blue Sage Dairy in 2010. Greenawalt makes both her own goat's milk cheeses as well as sheep's milk cheeses for Sluder at Sluder's on-farm facility. "The investment in starting a cheese plant is so huge," says Greenawalt. "There's no way I would have been able to start my own dairy without Laura's help." The two distinct cheese operations function generally harmoniously under one roof, sharing facilities, equipment, and, as it happens, a cheesemaker. Several of Oregon's cheesemakers are forging similar alliances. Mariano Battro works in conjunction with Francisco Ochoa, using Ochoa's facilities at Ochoca's Queseria in Albany, Oregon, to produce cow's milk cheeses for his own cheese company, La Mariposa. Likewise, Brian and Kate Humiston started Full Circle Creamery in 2010 by taking advantage of an existing organic milk supply and cheesemaking facilities at Noris Dairy near the small Willamette Valley town of Scio, Oregon. They're very happy with how things are working out. "We don't know how to be dairy farmers, " says Kate. "Making cheese, marketing cheese, and distributing cheese takes all of our time. We're more than happy to leave the dairy farming to the people who know what they're doing." Working with already-established operations, these cheesemakers are finding creative ways to make cheese while mitigating the steep up-front costs inherent in such an endeavor.

Another idea sometimes raised in artisan cheesemaking circles in the United States is the establishment of protected designations based on European models. For example, the French *appellation d'origine contrôlée* (AOC) designation, granted by the French government to producers of Brie de Meaux cheese, dictates a prescribed region where that style of cheese may be produced and the recipe by which it must be made. The AOC model and its equivalent in other European countries has had the effect of creating strong place-based brand identities for cheese, wine, and other products and imparts a significant marketing boost to those bearing the status. The Pacific Northwest landscape presents myriad opportunities for such a place-based designation model; few areas of the country are as richly diverse in both geology and geography. The marine climate along the Oregon and Washington coastline, the many fertile valleys including the Willamette Valley in Oregon and the Skagit Valley in Washington, the mountainous alpine regions of the Cascade Mountains and the Rockies of eastern Idaho all represent potent opportunities for cheese-makers looking to build an identity around a given area's *terroir*. The United States wine industry has already taken steps in this direction, establishing American Viticultural Area (AVA) designations in wine-producing regions across the country (including twenty-six in the Pacific Northwest), though an AVA designation is only indicative of a specific region of origin and does not go so far as to dictate production methods like the European model. At least for now, the domestic artisan cheesemaking community has not embraced the idea of such classifications for cheese but the concept holds promise for creating and nurturing a strong place-based identity that could raise the profile of the Pacific Northwest's cheeses and cheesemakers.

Perhaps what is most remarkable about the artisan cheesemaking community in the Pacific Northwest is that it has grown continuously, even despite the economic recession that spread across the nation in 2008. Among the newest operations in the Pacific Northwest is Mountain Lodge Farm in Eatonville, Washington, east of Tacoma, which was licensed to make cheese in early 2012. Owner and herdswoman Sherwin Ferguson has developed a mixed herd of La Mancha and Nigerian Dwarf goats and cheesemaker Meghan McKenna uses their milk to produce a variety of fresh and aged cheeses, including the lovely

Cheesemaker Meghan McKenna of Mountain Lodge Farm, Eatonville, Washington. Photo courtesy Sherwin Ferguson.

Summit, a bloomy-rinded, pyramid-shaped cheese that resembles nearby Mt. Rainier. Ferguson explained her desire to start an artisan cheesemaking operation, even in the face of all of the practical and financial obstacles, this way:

> The desire to use my land wisely and create wonderful healthy products with respect and love for these amazing animals are a big part of why I started Mountain Lodge Farm. A love of nature and the desire to try to make a positive difference in my little section of the world also drive me. I can't control so many things in the world, but I hope to make a small difference doing something positive, creative and honest . . . I do it because it makes me complete, fulfilled, and very much alive.

While for grateful consumers the artisan cheese movement's most visible product is its delicious cheeses, the Pacific Northwest's artisan cheesemakers are dedicated and passionate people who view their endeavors in the context of the broader importance of sustainable farming, land stewardship, and the production of wholesome, healthy food. This hardy and resourceful group contributes much more to the region's economy than just cheese. Their hard work and commitment will continue to sustain the environment, the community, and the domestic cheese movement for decades to come.

Glendale Creamery, Chimacum, Washington, date unknown. Photo courtesy Jefferson County Historical Society.

Appendix A
A Short History of Cheese in Alaska

Russian explorers first ventured east across the Pacific Ocean to the area we now call Alaska during the early eighteenth century, and the abundance of fur-bearing mammals they found there sparked commercial interest in the region. Russian fur trappers moved quickly to capitalize on the area's abundant natural resources and established a number of trade outposts in the region. But in order to survive in the remote and unfamiliar wilderness, the ambitious fur traders had to find a way to feed themselves. While the sea had sustained the Native populations for centuries, the Russians were not accustomed to a steady diet of fish and residents at the first Russian settlement on Kodiak Island kept a small herd of four cows and twelve goats. The animals supplemented the harvest from their small gardens, which yielded a few cabbages and potatoes.

In 1799, reigning Tsar Paul I granted the newly chartered Russian-American Company an operating monopoly in the region. The company turned the Kodiak Island settlement into a livestock-raising center and by 1818 as many as five hundred cattle, one hundred pigs, and one hundred sheep resided there. Most of the other Russian-American Company posts, including those at Unalaska Island in the Aleutian chain and at its headquarters at New Archangel (now the town of Sitka), also maintained a few cattle and cultivated small gardens. But the Russian-American Company quickly discovered that this style of organized agriculture was next to impossible in the region. Extreme climate conditions, including a very short growing season, bitter cold in the north and rainy, damp weather in the southern regions, prevented staples such as wheat and barley from ripening. The climate was also not very conducive to raising livestock, which did not do well in the harsh conditions and gave very little milk. Many cattle simply starved because the company could not produce and store enough hay to feed them. As a result of the many problems, Alexandr Baranov, Russian-American Company chief for several decades, was forced to purchase food and supplies from wherever he could find them. Baranov traded with a number of nations including the British,

Spanish, and Americans, and many enterprising merchants found financial gain in supplying the Russians with the grain and meat they desperately needed. Bostonian Nathan Winship even attempted to establish a trading post and settlement on the Columbia River near present-day Clatskanie, Oregon, to facilitate his company's burgeoning trade business with the Russians, though his efforts were unsuccessful.

The ongoing task of maintaining a steady flow of provisions to the region in 1812 led the Russian-American Company to establish Fort Ross in what is now coastal Sonoma County, California, territory then held by Spain. The farm-focused operation managed to produce and ship grain, meat, and dairy products including butter and cheese north to Alaska. At its peak during the late 1830s Fort Ross and several of its associated farms farther inland boasted a combined herd of over eighteen hundred livestock. Despite the company's efforts, however, agricultural efforts at Fort Ross were only marginally successful, due in part to insufficient labor and lack of farming experience, as well as (somewhat ironically) the damp, foggy coastal climate, which stunted crop growth. The Russian-American Company continued to require a substantial amount of supplies from outsiders and contracted with, among others, the Puget Sound Agricultural Company (the Hudson's Bay Company subsidiary) for grain, meat, and dairy products that were produced at the British company's more successful farming operations headquartered at Fort Vancouver along the Columbia River.

A number of factors, including the worldwide decline of the fur trade, the ongoing steep expense of maintaining its North American colonies, as well as a desire to keep the territory out of the hands of the British, influenced Russia to sell its Alaskan holdings to the United States in 1867. During the next several decades a number of missionaries, both Protestant and Catholic, made their way northward in an effort to convert the Native populations, but it was not until the Yukon Gold Rush at the turn of the twentieth century that white settlers began to populate Alaska in large numbers. Gold turned the otherwise forbidding tundra instantly attractive, drawing upwards of a hundred thousand people from all over the world during the boom years. The gold rush also provided a significant and welcome economic stimulus to the developing Pacific Northwest states of Oregon, Washington, and Idaho, and the farmers and dairymen of the region were more than happy to export their goods to "The Northland" for profit.

Many who traveled to Alaska seeking gold instead took advantage of the Homestead Act, which Congress extended to Alaska Territory in 1897, and settled permanently in the region. Turn-of-the-century observers spoke hopefully of Alaska's economic and agricultural potential, often comparing the area to northern Europe. "[The region] is in about the same latitude as parts of Finland and the Scandinavian Peninsula, which are celebrated for their butter and cheese, and the climate is fully as mild." Settlers imported cattle to the region by ship or brought them overland through British Columbia to the region, and while some kept a few animals for personal use, others seized the opportunity to supply the captive population of miners and their families with milk and dairy products. In 1901 the Valdez Dairy was already up and running in the town of the same name on the southern coast, producing milk and butter. Farther inland in Fairbanks, dairyman Charles Hinckley amassed a herd of fourteen cows in 1908 and Hinckley's Dairy quickly developed a reputation in the region for good quality milk, butter, and dutch cheese. By 1914 there were at least two additional dairies in the Fairbanks area, operated by the Young and Gould families. A number of dairies also sprung up in the area around Harrisburg (present-day Juneau), where gold was discovered in the 1880s. Dr. Frank Reynolds started what is likely the first dairy in the area in 1889, aptly named the Gold Creek Dairy, with a herd of about twenty cows. A 1917 news report called Simeonof Island on the south side of Bristol Bay the "dairy center of the Alaskan peninsula" because it was home to two herds of cattle, numbering eighty in all.

Despite the efforts of Alaska's earliest dairy entrepreneurs, however, production levels were just a fraction of what was then being produced in Washington, Oregon, and Idaho. In 1909, farmers produced a reported 6,773 pounds of butter and 600 pounds of cheese, which declined to just 3,507 pounds of butter and 365 pounds of cheese ten years later. The low numbers are due in part to the area's small population (just over sixty thousand as of 1900) as well as the multiple practical barriers to producing butter or cheese, among them climate, the high cost of importing animals and equipment, and the expense of constructing proper facilities. Lack of reliable transportation also plagued farmers until the Alaska Railroad was completed in the early 1920s, finally connecting the major population centers of the territory.

In an effort to encourage local agriculture, the United States Department of Agriculture established a number of experiment stations in Alaska beginning in 1898. Led by Dr. Charles Georgeson, scientists investigated what crops were best suited for the climate and growing conditions and experimented with animal husbandry and dairying. At the Kenai research station on Bristol Bay, researchers collected a small herd of Galloway cattle, a breed native to Scotland, that they hoped might be inherently well suited to the rigors of the Alaskan environment. Researchers produced fresh butter and cheese at the Kenai station "in order to ascertain if dairying can be made a success in that region when cattle are fed wholly on feed grown there." The cheeses were not described as being made in any specific style, but station cheesemakers wrapped their pressed cheeses in cloth and aged them for four to six weeks, turning the wheels periodically during the aging period.

Livestock research was later moved to Kodiak Island, an area that, as the Russians had already discovered, provided more open prairies for foraging and enabled the development of larger herds. By 1907, the Kodiak Station had accumulated a herd of forty-six cattle. In 1911 scientists were planning to resume the manufacture of butter and cheese, with an eye toward "develop[ing] the milking quality in the Galloway (breed) so that settlers can eventually be supplied with cows that meet their needs for milk and at the same time retain the hardiness, the rustling qualities and the feeding qualities of the Galloway." The eruption of nearby Mt. Katmai in 1912 threw a wrench in these plans, however, when it deposited eighteen inches of ash on the island, damaging the station and depriving the herd of forage. In order to avoid the high cost of purchasing hay and other feed and transporting it to Kodiak Island, the station's cattle were eventually shipped south to the state of Washington.

As was the case in the rest of the United States, commercial dairying in Alaska transitioned from small farms to increasingly larger commercial operations during later decades of the twentieth century. One of the largest of the new generation of dairies, Matanuska Maid, grew out of a Depression-era relief program. During the 1930s, the Federal Emergency Relief Administration developed a plan to establish a series of one hundred agricultural colonies across the United States, among them one in the Matanuska Valley, northeast of Anchorage. While other New Deal colonies were developed among existing populations, the Matanuska colony was created where none previously existed. After enduring a rigorous application and vetting process, two hundred and two families from Minnesota, Wisconsin, and Michigan made their way north to Alaska Territory by train and ship in 1935. Once they arrived, settlers drew lots for forty-acre plots of land in the vicinity of the present-day town of Palmer and set out to make their fortunes. While not all settlers were happy or successful in their subsequent farming endeavors and quite a few left, the colony did eventually turn the valley into what is still one of the state's most productive agricultural regions.

The Matanuska Valley Farmers Cooperative Association was formed in 1936 to market the agricultural products grown or produced by the valley colony's residents, which included a variety of vegetables, dairy products, sausages, and meats. In the early years the cooperative's dairy, Matanuska Maid, produced cottage cheese, butter, and fluid milk that it distributed in populated regions of the territory (Alaska did not became a state until 1959). Matanuska

Maid grew into one of the state's largest fluid milk processors, later expanding its offerings to include products such as ice cream and sour cream. Meanwhile, the many ads for Kraft, Darigold, and Tillamook brand cheeses appearing in area newspapers reveal that most cheese consumed by Alaskan consumers was imported from outside the state. Matanuska Maid began to flounder in the 1980s, due at least in part due to a precipitous decline in the number of dairy farms in the state, and eventually declared bankruptcy. The company's financial woes prompted the state to assume control in order to preserve the dairy industry in the region, but even with its support, Matanuska Maid continued to bleed money and the company closed for good in 2007.

The farmstead cheesemaking wave spreading across the rest of the nation reached Alaska in the early 2000s and several small farmstead cheesemakers started up in the state. Windsong Farms, a small cow's milk dairy operated by Gary and Carla Beu, and Cranberry Ridge Farm, a goat dairy operated by Rhonda and Matt Shaul, were both located in the Matanuska Valley. Both have since closed, however, citing high production costs; the Shauls relocated to Williamstown, New York, where they re-started Cranberry Ridge Farm and as of 2012 are making and selling farmstead cheese. A third operation in Palmer, Healing Acres Goat Dairy, operated by Margie Buchwalter, was open briefly but ceased production due to unrelated family health problems. The small-scale boomlet seemed to take agriculture officials by surprise, as Alaska was then the only state in the nation that did not have regulations on its books covering commercial cheesemaking operations. The state formally adopted cheese production regulations in January of 2011.

As of 2012 the only cheese producer operating in Alaska is Matanuska Creamery, started by dairyman Kyle Beus to fill the void left by the closing of Matanuska Maid Dairy. Matanuska Creamery financed itself from a variety of sources including federal loans as well as by selling "cheese futures"—in essence, patrons were encouraged to purchase cheese ahead of time, with the creamery promising to deliver the cheese once it was produced. The creamery managed to pre-sell 6.5 tons of its Alaska Gold Nugget Cheddar in this manner. Matanuska Creamery opened for business in 2008 and currently processes and distributes fluid milk as well as making cheese. According to news reports, the company has struggled with a number of challenges since its inception, including a lack of available milk. The long-term prospects for what is at present Alaska's sole commercial cheesemaking operation appear uncertain.

For over two hundred years, those brave souls who have attempted dairying and cheesemaking in Alaska have encountered an astounding number of challenges. Climate, a short growing season, lack of access to quality dairy animals, and prohibitively long distances to potential markets are among the issues that have plagued the area's agriculture industry for generations. As a result, throughout Alaska's history much of the cheese and dairy products, as well as other goods, consumed by the state's residents have been imported from the outside. Still, there may be room for small-scale cheesemaking operations in the state; in fact, these operations may be especially suited to thrive in Alaska's unique economy. The state's population grew 13 percent between 2000 and 2010, suggesting that the pool of potential consumers is on the rise. While small producers can hardly contemplate shipping cheese to out-of-state markets due to the prohibitive costs, all of the state's larger metropolitan areas including Anchorage, Fairbanks, Sitka, and Juneau hold seasonal farmers markets, providing sales venues for locally produced products. Prospective cheesemakers have another sales outlet in Formagio's, an artisan cheese and specialty food shop, which opened in Anchorage in 2010. Ultimately, small-scale cheese producers who are able to navigate the unique challenges of manufacture and sales in Alaska may yet be able to forge a path to sustained survival—but history shows it won't be easy.

Appendix B

Artisan Cheesemakers of the Pacific Northwest 2012

Oregon

Alsea Acres Goat Dairy (Alsea)

Ancient Heritage Dairy (Madras)

Briar Rose Creamery (Dundee)

Cada Dia Cheese Co. (Prineville)

Fairview Farm (Dallas)

Fern's Edge (Lowell)

Fraga Farm (Sweet Home)

Full Circle Creamery (Corvallis)

Goldin Artisan Goat Cheese (Molalla)

Juniper Grove Farm (Redmond)

La Mariposa (Corvallis)

Mama Terra Microcreamery (Williams)

New Moon Goat Dairy (Chiloquin)

Noris Dairy (Scio)

Oak Leaf Creamery (Grants Pass)

Ochoa's Queseria (Albany)

Oregon State University (Corvallis)

Pholia Farm (Rogue River)

Portland Creamery (Molalla)

Quail Run Creamery (Gaston)

Rivers Edge Chevre (Logsden)

Rogue Creamery (Central Point)

Tumalo Farms (Bend)

Wild Goose Farm (Portland)

Willamette Valley Cheese Co. (Salem)

Washington

Alpine Lakes Sheep Cheese (Leavenworth)
Appel Farms (Ferndale)
Back Country Creamery (Tacoma)
Beecher's Handmade Cheese (Seattle)
Black Sheep Creamery (Adna)
Blue Rose Dairy (Winlock)
Cascadia Creamery (Trout Lake)
Chattaroy Cheese Co. (Chattaroy)
Conway Family Farms (Camas)
Country Morning Farm (Othello)
Cozy Vale Creamery (Tenino)
Dee Creek Farm (Woodland)
El Michoacano (Marysville)
Fairaview Farm (Sequim)
Frisia Dairy & Creamery (Rochester)
Golden Glen Creamery (Bow)
Glendale Shepherd (Whidbey Island)
Gothberg Farms (Bow)
Grace Harbor Farms (Custer)
Heron Pond Farms (Spokane)
Herron Hill Dairy (Home)
Jacobs Creamery (Chehalis)
Kurtwood Farms (Vashon Island)
Larkhaven Farm (Tonasket)
Little Brown Farm (Whidbey Island)
L. J. Ranch (Ocean Park)
Meyers Goat Creamery (Orcas Island)
Monteillet Fromagerie (Dayton)
Mt. Townsend Creamery (Port Townsend)
Mountain Lodge Farm (Eatonville)
Mystery Bay Farm (Nordland)
Our Lady of the Rock (Shaw Island)
Pine Stump Farms (Omak)
Pleasant Valley Dairy (Ferndale)
Port Madison Farm (Bainbridge Island)
Quail Croft Farm (San Juan Island)
Queseria Bendita (Yakima)
Quillisascut Farmstead Cheese (Rice)

River Valley Ranch (Fall City)

Rosecrest Farm (Chehalis)

Samish Bay Cheese Co. (Bow)

St. John Monastery Farm (Goldendale)

Silver Springs Creamery (Lynden)

Steamboat Island Goat Farm (Olympia)

Sunny Pine Farm (Twisp)

Tieton Farm and Creamery (Tieton)

Twin Oaks Creamery (Chehalis)

Washington State University Creamery (Pullman)

Whiskey Hill Farm (Port Townsend)

Wild Harvest Creamery (Chimacum)

Willapa Hills Farmstead Cheese (Doty)

Yarmuth Farms (Darrington)

Idaho

Ballard Family Dairy and Cheese (Gooding)

Blue Sage Farm (Shoshone)

Green Goat Dairy (Shoshone)

Grumpy's Goat Shack (Victor)

Lark's Meadow Farm (Rexburg)

Manwaring Cheese Co. (Rigby)

Rollingstone Chevre (Parma)

Teton Valley Creamery (Driggs)

Wheyward Goat Cheese (Priest River)

Wood n' Goat Garden (Sterling)

Notes and Sources

Chapter 1

One of the less-discussed aspects of this era: Quote from John Boit, "John Boit's Log of the Columbia 1790-1793," *The Quarterly of the Oregon Historical Society* 22 no. 4 (1921): 274. The practice of carrying animals on ship continued well into the nineteenth century; see for example Anna Marie Pittman's letter written while traveling on board ship to Oregon in 1837: "I have heard the cock crow, the sheep bleat and the pig squeal. We have them on board, but few vegetables." Theressa Gay, ed., *Life and Letters of Mrs. Jason Lee* (Portland: Metropolitan Publishers, 1936), 126.

Of course "discovery" is a relative term: See discussion of Native American mortality in William G. Robbins, *Landscapes of Promise: The Oregon Story 1800-1940* (Seattle & London: University of Washington Press, 1997), 57-60.

As European fur-trading activity grew: Reports of the exact number of animals that arrived on the *Tonquin* vary. According to Robert Jones, ed., *Annals of Astoria: The Headquarters Log of the Pacific Fur Company on the Columbia River 1811-1815* (New York: Fordham University Press, 1999), 6, the vessel arrived with four goats, fourteen pigs, and two sheep.

For a variety of reasons, not the least of which: 1814: Barry Gough. ed., *The Journal of Alexander Henry the Younger 1799-1814 Vol. 2* (Toronto: Champlain Society, 1988-1992), 729, 693. Henry later mentions the presence of two goats, one that gave a quart of milk in the morning with sufficient milk left for a kid, and another that gave three pints of milk in the evening (712). Number of animals in 1817: Peter Corney, *Narrative of Several Trading Voyages from 1813 to 1818* (Honolulu: Thos. G. Thrum, 1896), 81.

As had been the case for the Pacific Fur Company: Astor's ship *Tonquin* was almost destroyed attempting to enter the mouth of the Columbia River; the Hudson's Bay Company itself lost a number of vessels including supply ships *William and Ann* (1829) and *Isabella* (1830) at the Columbia Bar.

The failure of both the Pacific Fur Company: Richard Somerset Mackie, *Trading Beyond the Mountains: The British Fur Trade on the Pacific 1793-1843* (Vancouver, British Columbia: UBC Press, 1997), 19 (map of Columbia Department of North West Company). Simpson quotes from Frederick Merk, ed., *Fur Trade and Empire: The Letters of Sir George Simpson* (Cambridge: Harvard University Press, 1931), 47, 50.

Simpson sold his superiors on farming: James R. Gibson, *Farming the Frontier: The Agricultural Opening of Oregon Country 1786-1846* (Seattle: University of Washington Press, 1985), 9-11. The Red River Settlement caused considerable controversy between HBC and the North West Company, which resented HBC's intrusion in the region. Conflict between the two companies eventually developed into a full-fledged battle in which twenty-one HBC men were killed.

One of the first practical steps Simpson took: Quote from Merk, *Fur Trade and Empire*, 124-25.

When John McLoughlin took up residence: 1825: John A. Hussey, "Fort Vancouver Farm" (unpublished manuscript, n.d., Collection of Fort Vancouver Library, Vancouver, Washington), 12. 1826: Gibson, *Farming the Frontier*, 38. Jason Lee: Jason Lee, "Diary of Rev. Jason Lee – II," *The Quarterly of the Oregon Historical Society* 17, no. 3 (1916): 262. An apple tree believed to have been among Fort Vancouver's original fruit trees still stands on the banks of the Columbia River at Old Apple Tree Park in Vancouver, Washington. Eventual extent of farm operations: Gibson, *Farming the Frontier,* 36.

All of what HBC termed "country produce": *Proceedings of the Alaska Boundary Tribunal* Vol. 3; (Washington D.C: US Government Printing Office, 1904), 209-14.

Following the agreement with the Russians: Gibson, *Farming the Frontier,* 90; British and American Joint Commission for the Final Settlement of the Claims of the Hudson's Bay and Puget's Sound Agricultural Companies, *Proceedings* vol. 3 (Montreal: J. Lovell, 1868), 105 (William Fraser Tolmie's description of Cowlitz Farm).

Livestock were a critical component: Source of cattle: In a letter, John McLoughlin notes, "in 1825, when I took charge of this place, I had only 3 bulls, 23 cows, 5 heifers and 9 steers." (E. E. Rich, ed., *The Letters of John McLoughlin from Fort Vancouver to the Governor and Committee First Series* [Toronto: Champlain Society, 1941], 207); see also Gibson, *Farming the Frontier,* 37 (source of additional cattle brought to the region). Fort Vancouver herd: Hudson's Bay Company Archives (HBCA) B.223/d/13:15 (1828 figures); Rich, ed., *The Letters of John McLoughlin First Series,* 207 (1837 figures). On Tualatin Plains: Inventory records show 129 cattle at the plains site in 1839 and 39 in 1840. See HBCA B.223/d/115:70 (1839); HBCA B.223/d/126:63 (1840). Cattle: HBCA B.223/d/155:81 and 88 (cattle numbers in 1844); for numbers of goats see, e.g., HBCA B.223/d/42:7 (goats at Fort Vancouver in 1832).

Most of HBC's regional outposts: Fort Colville: HBCA B.223/d/126:79. Fort Langley: HBCA B.223/d/155:81. Fort Nisqually: British and American Joint Commission, *Proceedings* 3:107 (testimony of William Fraser Tolmie).

As the earliest fur traders at Fort Astoria: McLoughlin quotes: Rich, *The Letters of John McLoughlin, First Series,* 44, 51, 207.

HBC established several formal dairies: Milk house: HBCA B.223/d/42:7. Narcissa Whitman quote: Clifford Drury, ed., *First White Women over the Rockies: Dairies, Letters and Biographical Sketches of Six Women of the Oregon Mission Who Made the Overland Journey in 1836 and 1838* (Glendale, California: A. H. Clark Co., 1963-66), 1:103. Four dairies: HBCA B.223/d/155:88. The name "Sauvie Island" is derived from the name of Laurent Sauvé, one of HBC's dairymen. Other milk houses: T. C. Elliott, "British Values in Oregon," *Oregon Historical Quarterly* 32 no. 1 (1931): 37 (Fort Hall); 39 (Fort Nez Percé); 40 (Forts Boise and Okanogan).

Their pans are of an oblong square: Drury, *First White Women over the Rockies,* 1:103.

After it was made, the butter was packed: Charles Wilkes, *Narrative of the United States Exploring Expedition During the Years 1838, 1839, 1840, 1841, 1842* (Philadelphia: Lea & Blanchard, 1845), 4:334. George Simpson, *Overland Journey Round the World During the Years 1841 and 1842* (Philadelphia: Lea & Blanchard, 1847), 106.

We know that the Hudson's Bay Company: Inventory: HBCA B.223/d/105b:48. Wilkes, *Narrative of the United States Exploring Expedition,* 4:307, 334. George Emmons, a member of Wilkes' expedition, also mentions dairy and butter production at Wapato Island; see "Extracts from the Emmons Journal," *The Quarterly of the Oregon Historical Society* 26 no. 3 (1925): 273. These accounts are echoed in an 1843 account of Fort Vancouver; see "Route Across the Rocky Mountains" (Overton Johnston and William Winter, *The Quarterly of the Oregon Historical Society* 7 no. 1 (1906): 62-104).

While no records have yet been uncovered: G. E. Fussell, *The English Dairy Farmer 1500-1900* (London: Frank Cass & Co. Ltd., 1960), 239-40.

Despite the relatively primitive state: Quantities: See, for example, Val Cheke, *The Story of Cheesemaking in Britain* (London: Routledge & Kegan Paul, 1959), 115, who notes that in 1730 London merchants received over five thousand tons of cheese from the Cheshire region and nearly as much from other parts of the country. See also Fussell, *The English Dairy Farmer,* Chapter 6: "The Trade in Dairy Products: Butter and Cheese." Categories: Fussell, *The English Dairy Farmer,* 244-45; a similar north/south division is also cited in Cheke, *The Story of Cheesemaking in Britain,* 123-24. Both Cheke and Fussell derive their conclusions in part from the work of English agricultural writer William Marshall, who toured the British countryside during the late eighteenth century and described cheesemaking practices, among other things, in his *Rural Economy* series of books.

In order to give the farming establishment: Rich, *The Letters of John McLoughlin, First Series,* 161 fn. 1.

Mrs. Capendale's position: Duties: Henry Stephens, *The Book of the Farm: Detailing the Labors of the Farmer, Steward, Ploughman, Hedger, Cattle-Man, Shepherd, Field Worker and Dairy-Maid* (Auburn: Alden Beardsley & Co., 1852), 1:162. See also Josiah Twamley, *Dairying Exemplified: or, the Business of Cheese-making,* (Warwick: J. Sharp, 1782). Cowlitz Farm: A. C. Anderson, "Memorandum Relating to Cowlitz Farm & c.," Mss 1502, Oregon Historical Society (pages unnumbered).

George Roberts, a clerk at Fort Vancouver: George Roberts, "Letters to Mrs. F. F. Victor," *Oregon Historical Quarterly* 63 no. 2 (1962): 225. Rich, *The Letters of John McLoughlin, First Series,* 161 (McLoughlin letter).

By November of 1836, McLoughlin: Rich, *The Letters of John McLoughlin, First Series,* 183. In "The Women of Fort Vancouver" (*Oregon Historical Quarterly* 93 no. 2 [1991]), John Hussey suggests that another reason for Mrs. Capendale's disaffection may have been her disapproval of so-called "country marriages" (not performed by clergy)which were common practice among fur traders. The marriage between John McLoughlin and his wife was also a country marriage, at least until he formalized it under pressure from the also-disapproving Reverend Beaver. Mrs. McLoughlin was also half Indian, another fact that may not have endeared her to Mrs. Capendale.

Some of the earliest non-HBC-affiliated: Carlos Schwantes, *The Pacific Northwest: An Interpretative History Revised & Enlarged Edition,* (Lincoln and London, University of Nebraska Press, 1989 & 1996, 96.

Regardless of their religious persuasion: Wascopam: Robert Boyd, *People of the Dalles: The Indians of the Wascopam Mission* (Lincoln and London: University of Nebraska Press, 1996), Appendix 1, Document 1 (Henry Perkins, "History of the Oregon Mission: Part One Wascopam Spring 1838"), 240. Waiilatpu: Clifford Drury, ed., *Marcus and Narcissa Whitman and the Opening of Old Oregon,* (Glendale, California: Arthur H. Clark Co., 1973), 1:238.

Cows provided milk and it fell: Pittman quote: Gay, ed. *Life and Letters of Mrs. Jason Lee,* 163. Simpson, *Overland Journey Round the World,* 104. Whitman quote: Drury, ed., *Marcus and Narcissa Whitman,* 1:421. Among the items noted by Henry Spalding in an 1847 inventory taken after the Whitmans were killed was one cheese press (2:336). Walker: Drury ed., *First White Women Over the Rockies*; see for example 2:269, 2:296, 2:314.

Lacking most of the comforts of home: Drury, *First White Women Over the Rockies,* 2:269.

The Catholic Church had established a strong presence: The Catholic Church operated as many as nineteen missions in the region between 1838 and 1870. See Wilfred P. Schoenberg, S.J., *A History of the Catholic Church in the Pacific Northwest 1743-1983* (Washington D.C., The Pastoral Press 1987), 201.

Like their Protestant counterparts: United States War Department, *Reports of Explorations and Surveys to Ascertain the Most Practicable and Economical Route for a Railroad from the Mississippi River to the Pacific Ocean 1853-5,* 36th Cong., 1st sess. Vol. 12, bk. 1, (Washington, D.C.: Thomas H. Ford Printer, 1860), 133 (Stevens quote); Hubert Howe Bancroft, *History of Washington, Idaho and Montana 1845-1889*; 603 (discussion of mission at St. Ignatius). The St. Ignatius mission was relocated to the Flathead Valley of Montana in 1854. See also William N. Bischoff, *The Jesuits in Old Oregon 1840-1940* (Caldwell, Idaho: Caxton Printers, 1945).

Inevitably, tensions developed between: John C. Jackson, "Red River Settlers vs. Puget Sound Agricultural Company," *Oregon Historical Quarterly* 95 no. 3 (1984): 278-79.

Though HBC's operations continued: Squatters: John A. Hussey, "The Fort Vancouver Farm," 146. Testimony published in the British and American Joint Commission *Proceedings* is full of accounts of squatters claiming company land, moving fences, shooting livestock, and generally harassing HBC employees.

Chapter 2

The Louisiana Purchase of 1803: Though Kelley had never even seen the region, in the throes of his enthusiasm he went so far as to plot out a settlement sited at the confluence of the Willamette and Columbia rivers, for which settlers were recruited. For his efforts, present-day Kelley Point Park in Portland, Oregon, at that very spot, is named in his honor.

By the mid-nineteenth century a substantial: Bailey quote: Sandra Myers ed., *Ho For California: Women's Overland Diaries from the Huntington Library* (San Marino: Huntington Library, 1980), 65. Migration numbers: John Unruh Jr., *The Plains Across: The Overland Emigrants and the Trans-Mississippi West 1840-1860* (Urbana and Chicago: University of Illinois Press, 1979), 395.

The physical rigors endured by animals: Young and old animals: Lansford Hastings, *The Emigrants' Guide to Oregon and California* (Princeton: Princeton University Press, 1932 [reprint]), 145. Adams quote: Cecilia Adams, "Crossing the Plain in 1852," *Transactions of the Oregon Pioneer Association* 1904, 323.

Despite the daily struggles of life: Hastings, *Emigrant's Guide*, 145. Reuben Gold Thwaites, ed. *Journal of Travels over the Rocky Mountains* by Joel Palmer (Cleveland: Arthur H. Clark Co., 1906), 258. Haun quote: Lillian Schlissel, ed., *Women's Diaries of the Westward Journey* (New York: Schocken Books, 2004), 172. See also Thomas D. Clark ed., *Gold Rush Diary: Being the Journal of Elisha Douglass Perkins on the Overland Trail in the Spring and Summer of 1849* (Lexington, University of Kentucky Press, 1967), 58 ("I was invited down to supper with them, where I had a treat of spending coffee with *milk*, hot biscuit & butter & c." [*emphasis in the original*]). Haun's reasons for traveling west are detailed in Sheila Rothman, *Living in the Shadow of Death: Tuberculosis and the Social Experience of Illness in American History* (New York: Basic Books, 1994), 134.

Those who managed to bring along a milk cow: Hendershott quote: Sanford and Sally Wilbur eds., *The McCully Train: Iowa to Oregon 1852* (Symbiosis Press, 2000), 100. According to the editors, Delilah Hendershott's account is unreliable in some respects, though these particular details are similar to accounts by other travelers. Baker quote: Kenneth L. Holmes, ed., *Covered Wagon Women: Diaries and Letters form the Western Trails 1840-1890* (Glendale, California: The Arthur H. Clark Co., 1984), 3:268; Carpenter quote: Myers, ed. *Ho for California*, 49. Carpenter quote: Myers ed., *Ho for California*, 149. It's also worth noting that most firsthand accounts of life on the trail that mention milk and/or butter are from the 1850s, when traveling had become considerably easier.

Given the severity of the Oregon Trail journey: Palmer quote: Thwaites, ed., *Journal of Travels over the Rocky Mountains* by Joel Palmer, 259. Carpenter quote: Myers ed., *Ho For California*, 119. Others also noticed and took advantage of the auto-churning technique: see, for example, Catherine Haun's account in *Women's Dairies of the Westward Journey* at 172 and "Sketch of Mrs. Susan P. Angell, a Pioneer of 1852" *Transactions of the Oregon Pioneer Association* 1928, 55. Adams quote: Cecilia Adams, "Crossing the Plain in 1852," 296.

Those not lucky enough to have their own cow: Dale L. Morgan ed., *The Overland Diary of James A. Pritchard from Kentucky to California in 1849* (Old West Publishing Co., 1959). Stories about the Laughlin family from William McNeal *History of Wasco County* (The Dalles: Bohn Printing Shop, 1952), 6 and 62; and Elizabeth Laughlin Lord *Reminiscences of Eastern Oregon* (Portland: The Irwin-Hodson Co., 1903), 78-95.

After shelter came the question of sustenance: Plimpton quote: Helen Betsy Abbott, "Life on the Lower Columbia 1853-1866" *Oregon Historical Quarterly* 83 no. 3 (1982): 264. Gilliam: Fred Lockley, "Reminiscences of Mrs. Frank Collins nee Martha Elizabeth Gilliam," *The Quarterly of the Oregon Historical Society* 17 no. 4 (1916): 358-72 (discussion of butter making on page 367).

Lucinda Collins' family was among the first: Information about Lucinda Collins Fares from Clarence Bagley, *History of King County, Washington* (Chicago and Seattle: S. J. Clarke Publishing Co., 1929), 782; essentially the same information can be found in Ada Hill, *A History of the Snoqualmie Valley* (1970), 15-17. Quote: Bagley, *History of King County*, 782.

Several early pioneers in the region: Bishop: James Hermanson, *Rural Jefferson County: Its Heritage and Maritime History* (Port Townsend: James Hermanson, 2002), 30. Davis: George

Lotzgessell, "Pioneer Days at Old Dungeness," *Washington Historical Quarterly*, 24 no. 4 (1933): 266-67; see also "The Pioneer Davis Family of Lopez Island, Dungeness," *Seattle Daily Times*, March 6, 1960.

Farmers who settled near more populated areas: Bryon: *Idaho Tri-Weekly Statesman*, July 11, 1874. Millman: *Idaho Triweekly Statesman*, June 8, 1876; a year earlier, Mrs. Millman had also sent cheese to the editors of the *Statesman*, which the paper also praised as being "as good as California cheese" (June 19, 1875).

Portland, the largest city in the Pacific Northwest: *Morning Oregonian*, August 26, 1861. The Ebenezer Quimby homestead was located in the vicinity of what is now the Parkrose neighborhood in northeast Portland. Dufur: *The Dalles Daily Chronicle*, June 8, 1892 (A. J. Dufur, Jr. making cheese at his ranch); additional information on Dufur family from Joseph Gaston, *Portland, Oregon: Its History and Builders* (Chicago and Portland: S. J. Clarke 1911), 3:792-96. Though he later returned to Portland, during this period A. J. Dufur, Jr., also platted out the Wasco County town that today bears the family name.

Astoria, at the mouth of the Columbia River: John Minto, "As Things Were in 1845," *Morning Oregonian*, February 1, 1904; Sarah Owens' experiences are outlined in daughter Bethenia Owens-Adair's *Dr. Owens-Adair: Some of Her Life Experiences*, (Portland: Mann & Beach, 1906 (?)), 141-162. Quote: Owens-Adair, *Dr. Owens-Adair*, 61. The Hobson family came over the Oregon Trail with the Owens family.

Those who settled in more remote: Warren Vaughn, *Till Broad Daylight, A History of Early Settlement in Oregon's Tillamook County* (Wallowa, Oregon: Bear Creek Press, 2004), 21,44–45, 53. An extended discussion of the construction of the *Morning Star* can be found in Vaughn, Book Three.

The problem of access to markets was often: Morgan: A. J. Splawn, *Kam-ia-Kin – The Last Hero of the Yakimas* (Portland: Kilham Stationery and Printing Co., 1917), 275. Kiester and the means of butter transport: "Dairy Cows Have Often Saved Farmers of Kittitas," *Ellensburg Record*, July 6, 1953. Stephens: Lois Barton, "Cheese Factory at Vaughn," *Lane County Historian* 28 no. 2 (1983): 51. Hansen: "Cheese Making Early Industry," *Twin Falls Daily News*, July 15, 1918. The town of Hansen, Idaho, east of Twin Falls, is named after John Hansen.

The many mining and lumber camps: Cynthia Applegate: Sallie Applegate Long, "Sketch of the Life and Character of Mrs. Jesse Applegate," n.d., Applegate Family Papers circa 1830-1940, Mss 233, Oregon Historical Society. Jesse Applegate: Sallie Applegate Long, "Mr. Jesse Applegate," n.d., Applegate Family Papers circa 1830-1940, Mss 233, Oregon Historical Society. Bennett: Paula Jo Connelly, "A Natural Dairy Producing Country: The SW Idaho Dairy Industry 1860-1940" (Masters Thesis, Boise State University, 2003), 17. Boswell: Claudia Killough, *Bits of Wallowa County Lore* (Enterprise, Oregon: Wallowa County Chieftain, 1971), 23. Larger companies such as the Potlatch Lumber Co. in Northern Idaho established company towns around their lumber and mining businesses and became large institutional purchasers of farm products including milk, butter, and cheese.

The production of dairy products increased steadily: Ninth United States Census, 1870 (Washington D.C.: Government Printing Office, 1872) gives state by state figures for dairy products production in 1850, 1860 and 1870.

Historically, the consumption of butter: Consumption numbers: U.S. Department of Agriculture, Economic Research Service, Food Availability (Per Capita) Data System, http://www.ers.usda.gov/Data/FoodConsumption. United States butter consumption peaked at an amazing 18.7 pounds per person per year in 1911. Alvord: Henry Alvord, "The Manufacture and Consumption of Cheese" in U. S. Department of Agriculture, *Yearbook of the United States Department of Agriculture 1895* (Washington D. C.: Government Printing Office, 1896).

As settlers continued to stream into the region: The journal of Cranston's first wife, Susan Amelia Cranston, is included in Kenneth L. Holmes, ed., *Covered Wagon Women: Diaries and Letters form the Western Trails 1840-1890 Vol. 3* (Glendale, California: The Arthur H. Clark Co., 1984). "Know Your City and State," *Morning Oregonian*, April 5, 1928, (questions answered by Warren Cranston's son C. K. Cranston). Local newspaper: *Daily Capital Journal*, May 29, 1895.

30,000 pounds: Oregon State Dairy and Food Commissioner, *First Biennial Report 1892-1894*, 51; see also "Cranston's Cheese Factory," *Willamette Farmer*, September 20, 1873. Republicanism: *Daily Capital Journal*, March 23, 1896.

Another Willamette Valley farmer: "Ankeny's Farm," *Willamette Farmer*, May 18, 1872; See also "Ankeny's Dairy Farm," *Willamette Farmer*, September 21, 1872.

By the 1880s a thousand cows: Information about Clatsop County dairying and cheesemaking in the later nineteenth century from: *Daily Astorian*, June 20, 1879; "Clatsop County Dairy Interests," *Morning Astorian*, September 19, 1885; "Clatsop Plains—Stock and Dairy Interests," *Daily Astorian*, July 12, 1888; "Clatsop County: Growth and Prosperity of Astoria and its Surroundings," *Morning Oregonian*, January 1, 1883; "Clatsop County Cheese," *Daily Morning Astorian*, September 25, 1885. West's cheese: "A Successful Dairyman," *Sunday Oregonian*, January 29, 1899. Ocean: "Clatsop County Cheese," *Daily Morning Astorian*, September 25, 1885.

Dairy farmers just across the Columbia River: Hathaway: *Weekly Oregonian*, October 12, 1872. Stiles: *Vancouver Independent*, April 13 1877 (notes regarding the "superior facility"). See also *Monthly Oregonian*, January 12, 1874; *The Daily Astorian*, September 26, 1877 ("there has been manufactured during five months and twenty days [at Stiles' farm] of this summer and fall 23, 179 pounds of cheese"). 63 cows: "Fern Prairie, Sunnyside, Washougal," *Vancouver Independent*, June 14, 1877. Cramm: "Brush Prairie Items," *Vancouver Independent*, August 12, 1876; See also May 20, 1876 ("Whitney and Co. were lining a cheese vat the other day for Mr. Cramm of Brush Prairie").

Henry Koch started a creamery: Early Woodland Cheese Factory history from: Donald E. McIntosh (son of cheesemaker Peter McIntosh), "A History of Cheese Making in Woodland, Washington and the Adjacent Country," *Cowlitz County Historical Museum Quarterly* 9 (1968): 25-28; Wendy Scott, "Dairymen Built Cheese Plant to Solve Problem," *Cowlitz County Historical Museum Quarterly* 45 (2003): 23-36; and Judy Card, "Fields of Flowers and Forests of Firs, A History of the Woodland Community 1850-1958" (revised 1999) accessed at http://lewisriver.com/jcard1.html. The McIntosh article states that Henry Koch started his factory in 1885 but other sources put the date at 1887.

One of the newly formed Woodland Dairy: The opening of the Woodland facility was reported in *Vancouver Independent*, July 24, 1889. McIntosh is often called the father of Tillamook cheese for his efforts in helping to establish a viable cheese industry in Tillamook. McIntosh went on to own and operate four cheese plants in the Tillamook area before going out of business in 1905. He later became a dairy inspector in Portland and then something of an itinerant cheesemaker, working at or consulting in the operation of cheese plants in Menlo, Silvana, and Battle Ground, Washington, as well as at Coos Bay and Coquille, Oregon. He started his last cheese factory at Gaston, Oregon, in 1926 and worked there until his death in 1940. See Dean Collins, *The Story of Tillamook: The Little County that Became the Big Cheese* (Portland, Oregon: *Oregon Journal*, 1933, reprinted 1961), Chapter VII.

Because the Pacific Northwest was growing: *Puget Sound Weekly Argus Supplement*, April 26, 1878. *Puget Sound Weekly Argus*, May 24, 1878 (shipping cheese), September 5, 1878 (next year's plans).

To the south in Lewis County: Reporter: *Washington Standard*, October 12, 1878 and May 16, 1879. See also *New Northwest*, May 15, 1879 (plant making three hundred pounds of cheese per day). Two other factories: Hubert Howe Bancroft, *The Works of Hubert Howe Bancroft Vol. XXXI: The History of Washington, Idaho and Montana 1845-1889* (San Francisco: The History Company, 1890), 344. See also *Washington Standard*, June 25, 1880 (White River plant making three to four hundred pounds of cheese/day). Birmingham later founded the William Birmingham Company in Tacoma, a wholesale and retail grain business. Hanna: McIntosh, "A History of Cheese Making in Woodland, Washington," 25.

Entrepreneur Thaddeus (T. S.) Townsend: Leland Townsend "The First Cheese Factory in Tillamook" *Oregon Historical Quarterly* 74 no. 3 (1973), 274-77. See also "Pioneer of Creameries," *Morning Oregonian*, January 1, 1904 (this latter article puts the date of the founding of the first creamery at Tillamook as 1891, though most other sources put the date at 1894).

Until the mining boom of the 1860s: Rich quote: Dean L. May, "Mormon Cooperatives in Paris, Idaho 1869-1896," *Idaho Yesterdays* 19 no. 2 (1975): 19. Swiss to cheddar: Russell Rich, *Land of the Sky-Blue Water: A History of the LDS Settlement of the Bear Lake Valley* (Provo, Utah: BYU Press, 1963), 123. See also "Co-operation in the Bear Lake Valley" *Deseret News*, March 21, 1877 and "Co-operative Industries," *Deseret News*, August 22, 1877. 1883: Rich, *Land of the Sky-Blue Water*, 124-25. See also Bancroft, *The History of Washington, Idaho and Montana 1845-1889*, 549 fn. 10.

The transition between settlement: Eighteenth-century production: Loyal Durand Jr., "The Migration of Cheese Manufacture in the United States," *Annals of the Association of American Geographers*, 42 no. 4 (1952), 264-66; Paul Kindstedt, *American Farmstead Cheese: The Complete Guide to Making and Selling Artisan Cheeses* (White River Junction, Vermont: Chelsea Green Publishing, 2005), 20-22; Paul Kindstedt, *Cheese and Culture, A History of Cheese and its Place in Western Civilization* (White River Junction, Vermont: Chelsea Green Publishing, 2005), Chapter 8, "The Puritans, the Factory and the Demise of Traditional Cheesemaking." Background on the development of cheesemaking in the United States can be found in Kindstedt, *American Farmstead Cheese*, Chapter 2; Kindstedt, *Cheese and Culture*; T. R. Pirtle, *History of the Dairy Industry*, (Chicago: Mojonnier Brothers Co., 1926); Durand Jr., "The Migration of Cheese Manufacture in the United States." 1890 production figures from U.S. Census 1890.

The Pacific Northwest dairy industry: *Morning Oregonian*, January 9, 1899.

Chapter 3

Across the nation, a new type of institution: "Moving Schools of Agriculture in the Inland Empire," *Spokane Spokesman-Review*, February 21, 1911.

In addition to their academic programs: Oregon demonstration train and slogan: "Train Route Fixed," *Morning Oregonian*, September 29, 1909. Demonstration boat: "Washington Instructors to Come Here," *San Juan Islander*, July 29, 1910 and "Farmer's Institutes," n.d., L. W. Hanson Papers 1905-1964, Washington State University. Farmers Institute events were announced frequently in local newspapers, which sometimes also listed topics and schedules. See, for example, "Farmers Institutes," *Olympia Weekly Capital*, September 4, 1903; "Farmers Institutes Open in City Tomorrow Morning, *Ellensburg Daily Record*, November 10, 1914 (complete program).

In 1900, Assistant Professor F. L. Kent: Kent's report on his creamery tour was published in Oregon Agricultural College's *Station Bulletin 65* (1901). "Teaching the Dairymen," *Morning Oregonian*, August 13, 1900, confirms that Kent made his tour by bicycle. See also John Burtner, "Twas a Great Event When Four State College 'Profs' 'Biked' to San Francisco and Back," *Oregon Journal*, April 24, 1938. The story of the trip back from San Francisco was particularly dramatic; at one point the men apparently decided to take a short cut through the Siskiyou Mountains, availing themselves of a railroad tunnel. Naturally, a train barreled its way through at precisely the same time and the men saved themselves by becoming one with the side of the tunnel. F. L. Kent went on to become a statistician for the Portland bureau of the United States Department of Agriculture.

Aside from his avocation, Professor Kent: Oregon Agricultural College, *Station Bulletin 65* (1901), 28. Kent lists seven additional facilities in Umatilla and Union counties (page 35).

Cheese production grew steadily: Thirteenth Biennial Report of Dairy and Food Commissioner, State of Oregon (Salem, Oregon: State Printing Department 1921), 23 (74 factories); United States Census of Agriculture, 1920 (state production numbers).

One reason for Oregon's growing dairy production numbers: Danish immigrant James Peter Morgan was also among the Tillamook area's first formal cheesemakers, making and selling his wares from his homestead near Kilchis Point during the 1860s (see Tillamook County Pioneer Museum file; Ada Orcutt, *Tillamook: Land of Many Waters* (Portland: Binfords & Mort 1951), 167. Morgan is said to have changed his name from Jens Peter Mortensen. According to Joseph Gaston, a Swiss-born cheesemaker, grandfather to Melchior Albplanalp, made swiss cheese in the Tillamook area during the 1880s (see Joseph Gaston, *A Centennial History* *of Oregon 1811-1912* (Chicago: S. J. Clarke and Co., 1912), 2: 854. Foland information from: *Oregon Journal* August 13, 1962; Collins *The Story of Tillamook*, chapter V. Note that Foland's name is often misspelled as "Folan" and "Folend."

The long distance between Tillamook: Collins, *The Story of Tillamook*, 148.

After several failed attempts at working cooperatively: Details on events leading up to the formation of the cooperative from Collins, *The Story of Tillamook*, chapters IX – XVI. Sources vary on the number of members that joined the original association. Collins says that ten joined (177). Reed Bain, "The Growth of an Institution: A Sociological Interpretation of Tillamook County Creamery Association of Tillamook, Oregon" (Ph.D. Dissertation, University of Michigan, 1926), 124, says that there were seven original subscribers. In its 1945 Annual Report the Tillamook County Creamery Association listed twelve original members. According to company representatives the original 1909 documents have not survived.

[T]o bring the patrons of the different creameries: U. S. Department of Agriculture, *Yearbook of United States Dept. of Agriculture 1917*, MacPherson, Hector, and W. H. Kerr, "A Federated Cooperative Cheese Manufacturing and Marketing Association in Tillamook County, Oregon" (Washington, D. C.: Government Printing Office, 1918).

In this earliest phase the TCCA's primary focus: Collins, *The Story of Tillamook*, 192-93.

Cheese production and sales increased: Botsford: "County Association History Traced Here," *Tillamook Headlight-Herald*, June 1, 1961. Printing on the rind: "Oldtimers Recall Early Tillamook Cheese Days," *Oregonian*, May 24, 1964. Production: Tillamook made 2,541,057 pounds of cheese in 1909 and 5,036,900 pounds in 1918. See also "Cheese Production in Tillamook Gains," *Morning Oregonian*, February 23, 1919.

While Tillamook's dairy farmers prospered: Rogers: See Emil R. Peterson and Alfred Powers, *A Century of Coos and Curry: History of Southwest Oregon* (Portland: Binfords and Mort, 1952), 341 (Rogers first cheesemaker in the county); *Coos River Sun*, March 31, 1892; "Wealth Made in Coos County"; *Morning Oregonian*, November 21, 1904 (lives of the Rogers family); Charlotte Mahaffy, *Coos River Echoes* (Portland: Interstate Press, 1965), 23-24. Moser: Background information from Joseph Gaston, *Centennial History of Oregon 1811-1912* (Chicago: S. Clark Publishing Co., 1912), 4: 817. In 1915, Andrew Christensen took control of Moser's factory at Gravel Ford, which he purchased in 1916 with an option on the North Fork Factory. See *Coos Bay Times*, June 25, 1915 (Christensen takes control over Moser creamery); *Coos Bay Times*, January, 2, 1916 (Christensen purchases). Harris: *Coast Mail*, June 15, 1901; "Coos Bay Product Winning More Honors," *Coast Mail*, September 20, 1902; "Milk from a Modern Dairy" *Coos Bay Times*, May 30, 1910.

The single largest dairy products factory: For more about the group's formation see Peterson and Powers, *A Century of Coos and Curry*, 341-42; "In Coos Bay Country," *Morning Oregonian*, May 4, 1900 (production numbers for 1899). Boat transport: See description in "More About Coos," *Bandon Recorder*, August 17, 1905.

The southern Oregon coast's growing reputation: *Semi-Weekly Bandon Recorder* July 3, 1914; "Italians Come to Bay," *The Coos Bay Times*, June 30, 1914; *The Coos Bay Times* August 19, 1914; "Vast Douglas County Tract Sought for Cheese Ranch," *Morning Oregonian*, August 16, 1914; "6000 Lane Acres Sell for $75,000," *Morning Oregonian*, August 2, 1914 (Griffin purchases land).

While the Coos Bay region possesses: McPherson quote: "To Raise Profits," *Coos Bay Times*, February 21, 1916. CCCCA: "Cows Make Money," *Coos Bay Times*, May 9, 1916; "Cheese Rules Made," *Morning Oregonian* May 7, 1916. According to the May 5 edition of the *Coos Bay Times*, of all of the cheese factories in the region only the Coos Bay Creamery and one cheese factory located in the town of Norway did not join the cooperative. Merger: Bain, "The Growth of an Institution," 319. According to Bain, a number of farmers' groups from both Oregon and Washington approached the TCCA about possible mergers but none were ever realized. "Cheese Output Valued at $800,000," *Morning Oregonian*, January 1, 1917. Testing and standardization: P. S. Lucas, "Problems in the Cheese Industry," *Elgin Dairy Report*, March 22, 1919, continued in March 29 edition, (survey taken of testing and standardization methods in the state; only Tillamook and Coos-Curry had developed programs).

Given the remoteness of Oregon's coastal regions: "Many Leave on *Redondo*," *Coos Bay Times*, Nov 8, 1913. "*Iaqua* Sails South Sunday," *Coos Bay Times*, August 18, 1913. "Launch Sinks With Cheese," *Coos Bay Times*, January 18, 1913.

By 1914 Brown had developed: Details of Brown Farm operations from: Mark Nielsen, "The Brown Farm on the Nisqually Delta, 1904-1919: A Photographic Essay," *The Pacific Northwest Quarterly* 71, no. 4 (1980) 162-71; A. Brown, "How I Reduced the Cost of Marketing," *System on the Farm* 1, no. 5 (1917), 175-78; "Story of $500,000 Farm That's Made Parcel Post Famous," *Tacoma Times*, March 7, 1918; "How a Bum Lawyer Became a Farm Expert," *Tacoma Times*, March 8, 1916; "Farmer Brown Didn't Spent a Fortune on Clearing His Lands," *Tacoma Times*, March 11, 1916. Ads for Brown Farm products appeared regularly in *The Seattle Daily Times* from 1914 to 1916.

One of the hallmarks: See "Notice of Trustee Sale," *The Seattle Daily Times*, July 21, 1918; *In re Brown*, 251 F. 365 (West. Dist. Wa. 1918). There seems to have been some intrigue, at least in the press, surrounding the disposition of Brown Farm debts and A. L. Brown's father's estate changed hands, possibly under mysterious circumstances, after settling all of the farm's debts. See "Mystery Veils Sale of Amos Brown Estate," *The Seattle Daily Times*, August 30, 1919.

While Oregon cheese production thrived: Condensed milk: Washington State Dairy and Food Commissioner, *8th Biennial Report 1910*, 34; Oregon State Dairy and Food Commissioner, *Biennial Report 1909*, 50. The Oregon Dairy and Food Commissioner reported seven condensaries in Oregon in his 1911 report. Meade: *Northwest Horticulturist*, August 1899, 191. See also "The Thriving Town of Kent," *Seattle Daily Times*, May 12, 1900 (quality); *Seattle Daily Times*, July 22, 1901 (closure); Meade's factory later reopened and operated for several more years before closing entirely.

Because of the active condensary industry: Cheney Creamery started: *Northwest Tribune*, May 3, 1890. Largest in state: *Spokane Chronicle*, July 3, 1897. Sold: "Cheney's Big Creamery is Sold," *Spokane Spokesman-Review*, April 16, 1899. Reuter: Background from "A Guide to the Cheney Central Historical District" and "Frederick Wilhelm Reuter," collection Cheney Historical Museum, Cheney, Washington.

Hazelwood was one of the Inland Empire's: "First Hazelwood Farm Located in Illinois," *Spokane Spokesman-Review*, April 16, 1933. *Northwest Tribune*, June 2, 1892. For a time Hazelwood also ran the creamery operations at the University of Idaho in Moscow: "Hazelwood Buys 'U' Dairy Output," *Spokane Spokesman-Review*, June 8, 1915. Hazelwood's musical events were announced regularly in the *Oregonian*; see, for example, "Special Music," *Morning Oregonian*, May 5, 1908. Two factories: "Cheese Output is Huge," *Morning Oregonian*, February 27, 1910. Western Dairy Products: Clarence Bagley, *History of King County, Washington* (Chicago and Seattle: S. J. Clarke and Co. 1929), 3:825 (biography of R. E. Campbell, one of the men behind the founding of Western Dairy products). By early 1927 Hazelwood's wholesale business was subsumed by Western as well.

The dairy story of the Kittitas Valley: *History of Klickitat, Yakima and Kittitas Counties*, 795. "Wholesale Houses That Make Seattle Their Headquarters," *Seattle Daily Times*, February 7, 1904.

Around the turn of the twentieth century: Kittitas acquires Cloverdale: *History of Klickitat, Yakima and Kittitas Counties*, 818. Goodwin's injury: "Creamery Notes," *Northwest Horticulturist, Agriculturist and Dairyman*, no. 9 September 1898, 137. Consolidated: *Ellensburg Daily Record*, January 22, 1915; see also "Old Creamery Adds Improved Methods," *Ellensburg Capital*, January 28, 1915.

Eventually larger-scale irrigation projects: Mark Fiege, *Irrigated Eden: The Making of an Agricultural Landscape in the American West* (Seattle and London: University of Washington Press, 1999), 16.

While water did not by itself create: Fiege, *Irrigated Eden*, 147; Carpenter quote *The Ranch*, July 7, 1894.

Convincing area farmers to look beyond: Info on Carpenter from *The Ranch*, April 14, 1894; July 7, 1894 (quote); *Yakima Herald*, October 17, 1895; *Yakima Herald*, March 10, 1898. Another Yakima area entrepreneur, Elizabeth Carmichael, started the region's well-known

Yakima City Creamery in 1902, which produced cheese, butter, cream, and ice cream. The Maid o' Clover chain of convenience stores, still in existence today, evolved out of the Yakima City Creamery operations in the 1960s.

Idaho's turn-of-the-twentieth-century dairy farms: "Cheese Factories Succeed on the Minidoka Project," *Reclamation Record,* November-December 1923, 319. During World War II the area became home to the Minidoka War Relocation Center that housed thousands of Japanese Americans.

Sixty miles or so to the west in Twin Falls: Oral History, Joseph Maxwell, O865, collection Idaho Historical Society. Oral History, Ted Sandemeyer, O867a-b, collection Idaho Historical Society. According to his account, the Sandemeyer family continued to operate the plant for several years before shutting it down. Area dairy farmers then attempted to organize, unsuccessfully, a cooperative to operate the factory. Kaeser: "New Cheese Factory," *Twin Falls Times,* January 20, 1914; the factory closed after World War I, though Kaeser attempted to revive it a few years later.

The Mormon settlements in the Bear Lake Valley: Factory numbers: "Southeastern Idaho," *Salt Lake Herald,* December 26, 1897.

Among the most prolific of the area's dairy: Additional information on the Kunz family from: Dan Kunz Oral History OH833; Denzil Kunz Oral History, OH832; Jack Kunz Oral History OH831, collection Idaho Historical Society. Production: "Bear Lake Jottings," *Deseret News,* August 5, 1885.

As farm production expanded across the Pacific Northwest: See, e.g., Oregon Agricultural College, *The Business Side of Farming,* by John Andrew Bexell and Hector MacPherson (Corvallis: Oregon Agricultural Press 1911), Part II Chapter 2, "Marketing Farm Products" for an in-depth description of the contemporary process for marketing farm products.

A farmer's arrangement with a wholesale merchant: Information on 1901 dispute from "Dispute Over Cheese: Makers and Dealers Fail to Agree," *Morning Oregonian,* June 3, 1901; "Richest in the State," *Morning Oregonian,* June 20, 1901; "Cheese For the Orient," *Morning Oregonian,* July 3, 1901 (merchant follows through on threats). Information on 1914 dispute from Bain, "The Growth of an Institution," 314-15.

In other cases, corrupt wholesalers: Fabricated stories: "Bogus Commission Merchants," *Ranche and Range,* July 18, 1901. Seattle: Alice Shorett and Murray Morgan, *Soul of the City: The Pike Place Public Market* (Seattle: The Market Foundation and University of Washington Press, 2007), 16. Childs: "Cheese Manufacturer Loses Large Sum to 'Frisco Swindlers'," *Morning Oregonian,* January 23, 1900.

Stark power inequalities between farmers and merchants: Seattle: "Public Market is Popular," *Seattle Daily Times,* August 17, 1907. Portland: James Beard, *Delights and Prejudices* (New York: Atheneum Press, 1964), 48-49. Boise: *Idaho Statesman,* August 1, 1915.

Though farmers markets were very popular: U. S. Patent No. 1,242,872 (filed October 21, 1916). Wholesalers: William H. Nicholls, "Post-War Concentration in the Cheese industry," *Journal of Political Economy* 47 no. 6 (1939): 826-27. Nicholls notes that another factor that contributed to the decline of wholesalers was the increasing popularity of processed cheese. Processed cheese manufacturers purchased cheese directly from factories, bypassing the wholesale distribution chain altogether.

The Pacific Northwest received a big economic boost: Carlos Schwantes, *The Pacific Northwest: An Interpretive History* (Lincoln and London: University of Nebraska Press, Revised and Enlarged Edition, 1996), 358.

Farm production soared across the United States: See United States Food Administration consumer pamphlet, accessible at: http://www.ecommcode.com/hoover/hooveronline/hoover_bio/archive/food/wheat.htm. *Morning Oregonian,* October 19, 1918 (Tillamook); *Seattle Daily Times,* June 21, 1918 (Kristoferson's).

Another important impact of World War I: "Shortage of Rennet is Serious," *Morning Oregonian,* June 25, 1916. Oregon Agricultural College Experiment Station, *The Use of Pepsin as a Rennet*

Substitute in Cheddar Cheesemaking, Bulletin 155 (1918). See also "Substitute for Rennet," *Morning Oregonian,* September 17, 1916.

Because milk is inherently perishable: This is a very brief summary of the enormously complex history of the fluid milk industry in the United States. For a more complete discussion of the many issues that led to the rise in demand for fresh milk see: E. Melanie DuPuis, *Nature's Perfect Food* (New York: NYU Press, 2002); Anne Mendelsohn, *Milk: The Surprising Story of Milk Through the Ages* (New York: Knopf, 2008); Sarah Valenze, *Milk: A Local and Global History* (New Haven and London: Yale University Press, 2011); Richard Meckel, *Save the Babies, American Public Health Reform and the Prevention of Infant Mortality 1850-1929* (Ann Arbor: University of Michigan Press, 1998).

The rapid growth in fluid milk consumption: Bureau of the Census, *Mortality Statistics 1900-1904,* Special Reports (Washington, D. C.: Government Printing Office, 1906); summary tables for leading causes of disease-related death page xxi, tuberculosis mortality tables page xxxiv.

German physician Robert Koch: Infection rate: Alan Olmstead and Paul Rhode, "An Impossible Undertaking: The Eradication of Bovine Tuberculosis in the United States," *Journal of Economic History* 64 no. 3 (2004): 33-34; estimates of the rate of bovine tuberculosis infection in humans varies between 10 and 30 percent of overall infections. Eradication: Olmstead and Rhode, "An Impossible Undertaking," 2-4. European countries took a much less radical approach to the issue of bovine TB, at least in part because of concern that the mass eradication of cattle could lead to food shortages. In Denmark, for example, only obviously sick or diseased animals were destroyed. The problem persists. In 2005, New York health officials traced a tuberculosis outbreak to contaminated cheeses imported from Mexico. At least thirty-five people became infected and one infant died from the disease. See Marc Santora, "Tuberculosis Cases Prompt Warning on Raw Milk Cheese," *New York Times,* March 16, 2005.

By the dawn of the twentieth century: Breeds: See Irmagarde Richards, *Modern Milk Goats* (Philadelphia: J. B. Lippincott, 1921), Chapter II: "Imported Milk Goats." Health: See, for example, Myrtle Meyer Eldred, "Goat's Milk May Solve Diet Problems," *Morning Oregonian.* April 15, 1936; Mrs. A. L. S. Hansen, "The Magic in Goat's Milk," *Dairy Goat Journal,* March 1933 (makes many incredible claims for the benefits of goat's milk, citing examples of a six-year-old boy cured of epileptic fits and a little girl able to take off her leg braces). See also *The Ranch,* July 15, 1905, touting a number of benefits of goats and their milk. TB free: See, for example, Richards *Modern Milk Goats,* 21 (goat's milk as "an ally against tuberculosis"); "Gentlemen, the Goat!" *Spokane Spokesman-Review,* December 20, 1910 ("Unlike the cow the goat has no inherited tendency toward tuberculosis."). Scientists now understand that goats can be infected by the same tuberculosis bacterium as cattle. *Oregonian* headline: *Morning Oregonian,* November 3, 1917.

Goat's milk became popular enough: *Seattle Daily Times,* September 27, 1920; "Milk Goat Body Meets," *Morning Oregonian,* October 22, 1920. *Seattle Daily Times,* October 31, 1920. Cook County Hospital ran a fairly extensive farm operation; see "Cook County Hospital Nets $14,000 in a Single Year," *Modern Hospital* XVIII no. 4 (1922): 318. See also *The Leavenworth Echo,* March 24, 1911 (announcing a goat farm in Meadowdale, Washington, in the Puget Sound region near Bremerton).

 Despite the proliferation of goat farms: Eggers: *Seattle Daily Times,* October 31, 1920. Williams: *Seattle Daily Times,* April 3, 1921.

Chapter 4

The World War I economic boom: See "Solution of Future is Oregon's Problem," *Morning Oregonian,* January 9, 1919 (Reconstruction Convention) and "Governor Says Oregon's Problems Without Precedent," *Morning Oregonian,* January 15, 1919; "Reconstruction Congress Speedy," *Seattle Daily Times,* January 10, 1919.

The postwar years also ushered in: USDA - National Agricultural Statistics Service, Total US Annual Cheese Production 1919-2010. U.S. cheese production was 479,390,000 pounds in 1919 and 417,895,000 pounds in 1920.

Postwar economic instability contributed: United States Department of Agriculture Rural Business Cooperative Service, *Co-ops 101: An Introduction to Cooperatives,* Cooperative Information Report 55 (1997), 8; dairy figures from United States Department of Agriculture, *Cooperatives in the Dairy Industry,* Cooperative Information Report 1 Section 16 (2005), 23. Tax exemption: Legally organized agricultural cooperatives were entirely tax exempt until 1951, after which legislation made taxation issues considerably more complicated. See Donald Frederick, "Income Tax Treatment of Cooperatives: Background," Cooperative Information Report Part I, 2005 Edition, United States Department of Agriculture.

Another style of cooperative evolved to facilitate: "Oregon Co-Operative Dairy Exchange Starts Business," *The Oregon Countryman* IX no. 1 (1916). Archie Satterfield, *The Darigold Story: The History of a Dairy Cooperative in the Pacific Northwest* (Seattle: Darigold Inc., 1993), 3-8.

While many cooperatives were successful: Emil Youngquist, Oral History RIII 156, collection Skagit County Historical Museum. Mt. Angel Cooperative background from Sister Joeine Darrington, OSB, "The Mt. Angel Cooperative Creamery: A Study of the Primary Factors Responsible for its Present Status" (Thesis, Dept. of Business Administration, University of Oregon, 1941); "Twenty Five Years of Cooperative Service: Mt. Angel Cooperative Creamery 1912-1937" (n.d., collection of Mt. Angel Historical Society).

One infamous regional cooperative: "Sacrifice of Cows Cutting Off Milk," *Morning Oregonian,* November 25, 1917.

Portland area milk distributors challenged: "Trust Law Broken," *Morning Oregonian,* January 5, 1918. "Milk War Starts," *Morning Oregonian,* January 24, 1918. "Dairymen Buy Up Plants," *Morning Oregonian,* July 1, 1920; "Alma Katz Resigns," *Morning Oregonian,* April 3, 1921.

Soon member farmers began to rebel: "Dairy League is Sued," *Morning Oregonian,* September 24, 1920. "510 members Quit Dairymen's League," *Morning Oregonian,* November 1, 1921. On the Nestlé Condensary, see Joe R. Blakeley, "The Nestlé Condensary in Bandon," *Oregon Historical Quarterly* 104 no. 4 (2003): 566-77. "Dairymen's League Votes to Disband," *Morning Oregonian,* December 28, 1921. For additional information on the rise and fall of the Oregon Dairymen's League, see Sharon Quinten Hoober, *Trends in the Integration of Cooperative Dairy Products Marketing Channels in the Pacific Northwest* (Masters Thesis, Oregon State College, 1949).

J. L. Kraft went on to revolutionize the cheese industry: U. S. Patent No. 1,186,524 (filed March 26, 1916). William H. Nicholls, "Post War Concentration in the Cheese Industry," *The Journal of Political Economy* 47 no. 6 (1939): 824-25.

With national demand for cheese increasing: J. L. Kraft, "Idaho's Dairying Future," *Cheesekraft,* August 1922, 5; Collection Kraft Archives. Visit to Idaho: J. L. Kraft notes in *Cheesekraft* that he was invited by the Boise Chamber of Commerce, suggesting that Governor Davis invited him only after the chamber had done so; see "Opening Address," *Cheesekraft,* March 1923, 4-5, Collection Kraft Co. Archives (J. L. Kraft address to Kraft Sales Convention) and "Speaks for Itself," *Cheesekraft,* March 1922, 2, Collection Kraft Co. Archives (text of invitation from Boise Chamber). The trip was also recounted widely in the local press. For a day-by-day, almost moment-by-moment, recounting of events, see "Easterners 'Discover' Idaho's Possibilities," *Idaho Farmer,* August 17, 1922. Rotary Club speech: "Cheese Maker Praises Idaho," *Twin Falls Weekly News,* June 1, 1922. See also "A 'Wire' From Idaho," *Butter Cheese and Egg Journal,* August 16, 1922. Production numbers: University of Idaho Agricultural Experiment Station, *The Dairy Situation in Idaho: Being Part II of a Tentative Report of the Agricultural Situation Based on An Economic Survey of the Production and Marketing of Idaho Farm Products,* Bulletin No. 152 (1927), 66.

Kraft was innovative in its methods: "Plans Announced for New Cheese Factories," *Twin Falls Daily News,* December 30, 1922; the Laabs established four factories in and around the Minidoka Project at Rupert, Paul, Burley, and Declo; see "Cheese Factories Succeed on the Minidoka Project," *Reclamation Record,* November-December 1923, 319. "Idaho is now Fifth State in Cheese Production," *Idaho Farmer,* October 7, 1926 (all of Laabs output sent to Kraft Pocatello plant; also lists all Laabs plants by city); "Kraft Buys Cheese Plant," *Idaho Citizen,* June 10, 1927. Nelson-Ricks, Mutual: State of Idaho Department of Agriculture, *Biennial Report* (1925-26), 18-19.

Like many early twentieth-century housewives: Doug Swanson, "Cheese Made Red Rock Roll," *Tigard Times,* February 1, 1978.

Red Rock Cheese Company was formally established: The sale of Red Rock to Kraft was significant enough to make the front page of the *Morning Oregonian* (September 24, 1929).

Clearly Kraft was focused on growth: "Kraft-Phenix Buys Ten Companies," *New York Times,* October 17, 1929. See, e.g., "Kraft Box, Shooks Plant will Move to Oregon City," *Morning Oregonian,* February 2, 1933; "Americans Eat More Cheese," *Oregonian,* October 20, 1939 (Kraft box plants in Bridal Veil and Cathlamet, Washington); "New Headquarters," *Seattle Daily Times,* December 2, 1936 (Seattle distribution warehouse established).

The increasing availability of goat's milk: *Salt Lake Tribune,* August 13, 1914 (Utah goat cheese makers); "Goat Cheese Factory in Humbolt," *Pacific Rural Press,* Feb 25, 1922. Oregon: "Milk Goat Body Meets," *Morning Oregonian,* October 22, 1920 (organization of Oregon Milk Goat Association); "Goat Dairy Stock Sold," *Morning Oregonian,* January 6, 1922 ("A small plant for making cheese . . . has been installed . . . and it is expected that commercial cheesemaking will start in the spring"). Portland Goat Dairy was located at the present-day intersection of NE Fremont and NE 82nd Avenue. Washington: *Butter, Cheese and Egg Journal,* November 16, 1921, 44.

Winnie and Jay Branson started one: Information on the Branson goat farm in Falls City from: *Morning Oregonian,* October 4, 1925 ("Oregon's First Roquefort Cheese"); May 23, 1926 ("Roquefort Cheese-Making Oregon's Newest Industry"); September 11, 1927 ("Roquefort Cheese Factory Thrives"); May 17, 1936 ("Saanen Kid Goes to New Bedford"); and C. N. Atwood, "Oregon Cheese Manufacture," *International Dairy Goat Journal,* August 1929, 4.

M. P. Eggers, one of Washington's dairy goat: Ad appeared in *Dairy Goat Journal,* April 1934. Eggers placed ads in goat publications around the country; see, for example, ad for Briar Hills Dairies in *New England Goat News,* January 1941.

Another regional goat cheese factory: Information on the Western Goat Products Company from: "Cheese Factory Incorporated," *Stanwood News,* January 12, 1928; "Cheese Factory to Start Soon," *Stanwood News,* March 22, 1928; *Seattle Daily Times,* July 8, 1928; "Local Cheese Factory Displays Products," *Stanwood Times,* November 22, 1928; *Seattle Daily Times,* September 27, 1928; on the issue of canvassing for goat's milk see, for example, "Cheese Factory to Start Soon," *Stanwood Times,* March 22, 1928 ("Mr. Clark has been called to address many commercial clubs in outlying towns and all are desirous of giving information for ranchers on the breeding of milk goats") and "Manufacturing New Product," *Stanwood Times,* April 12, 1928 ("Mr. Clark made a special trip to Arlington and gave the commercial club a very instructive talk on the raising of goats and told of the manufacturing of the product").

In 1929 George Munson purchased Western Goat: "Cheese from Goat's Milk; Stanwood Cheese Factory Now in the Hands of Experts," *Stanwood Times,* May 2, 1929.

In the paper *Fungi in Cheese Ripening*: United States Department of Agriculture Bureau of Animal Industry, *Fungi in Cheese Ripening: Camembert and Roquefort,* Bulletin No. 82 (1906). Quote from page 29. See also Hearings Before the House Committee on Agriculture, House of Representatives, 66th Congress, Second Session, December 9, 1919; see discussion 275-77; "Roquefort Made Here," *New York Times,* September 18, 1921; "Yankee Cheese is Being Shipped to Switzerland," *Seattle Daily Times,* August 14, 1920; "Acceptable Roquefort Cheese Made in United States From Cow's Milk," *Butter Cheese and Egg Journal,* Sept 28, 1921, 39. A photo of the foil-wrapped blue cheese produced by the Grove City Creamery can be found in T. R. Pirtle, *History of the Dairy Industry* (Chicago, Illinois: Mojonnier Brothers Co., 1926), 144.

Across the nation scientists began to apply: USDA Bureau of Agricultural Economics, Milk Production on Farms and Statistics of Dairy Products Plants 1950 (Feb 15, 1951).

Curry County, Oregon, in the far southwestern part: Info on Catterlin and the Star Ranch from "Dairy Farms Loom," *Morning Oregonian,* September 4, 1911 and "In Langlois Dairies Vast Riches Found," *Morning Oregonian,* November 24, 1912. Cooperative: "Langlois, Dairy Center of Curry County," *Morning Oregonian,* July 26, 1922. Information about Moore from: Ella Sponaugle, *Pistol River Recollections: A Series of Interviews* (Gold Beach, Oregon: Curry County Historical Society 2003), 155-56.

Danish immigrant Hans Hansen arrived: "Two Business Changes," *Bandon Recorder*, May 4, 1915 (Schmidt Hansen's businesses); "Weatherbee Quits Star Ranch," *Bandon Recorder*, Oct 12, 1915 (Hansen assumes lease of Star Ranch; brother Schmidt is owner of local shooting gallery); "Cheese Factory Rebuilding Eyed," *Oregonian*, January 28, 1958 (Hansen arrived in US from Denmark in 1915). Hansen buys ranch: "Oregon's Development Seen in Various Lines of Industry," *Morning Oregonian*, February 9, 1926. The Star Ranch changed hands several times before Hansen came on the scene, see, for example "Boosts Dairying," *Coos Bay Times*, December 19, 1913 (Dr. J. R. Wetherbee of Portland acquires the ranch). Reviewer: Fred Beck, "Farmer's Market Bulletin," *Los Angeles Times*, July 7, 1941. See also *Los Angeles Times* October 5, 1941 (ad for May Company Modern Market featuring Langlois cheese). Reporter: "Blue Star Cheese: 'Going to Town' at Langlois, Ore.," *National Butter and Cheese Journal*, December 1941, 13.

By the 1950s Langlois Blue was available: "News of Food," *New York Times*, January 25, 1952; "Food News: Coffee, Cheese" *New York Times*, April 25, 1955. According to the *Eugene Register-Guard*, Sven Knudsen, Danish-born cheesemaker at the nearby Reedsport Cheese Factory, was apparently also experimenting with a roquefort style cheese made from goat's milk. See "Reedsport Company Ships Much Cheese to Oregon Points," September 24, 1937. Mail order: "Blue Cheese Plant Burns," *Eugene Register-Guard*, April 28, 1957.

Tiny Langlois' cheesemaking fortunes: "Blaze Takes Cheese Plant," *Morning Oregonian*, April 27, 1957 (front page). Some remarkable photos of the plant as it is burning are reproduced in *The Langlois Centennial Committee Presents: The Pioneers of Langlois* (Langlois, Oregon: Langlois Centennial Committee 1981). New company: *Eugene Register-Guard*, November 21, 1968.

Klickitat County's earliest settlers: Info on Joseph Aerni, Wade Dean, and early Klickitat County dairying from Penny and Bill Collier, *Along the Mt. Adams Trail* (Hood River, Oregon: Maranatha Press, 1979), 164-77 and Keith McCoy, *The Mount Adams Country: Forgotten Corner of the Columbia River Gorge* (White Salmon, Washington: Pahto Publications, 1987), 91-95.

Natural caves are plentiful in this mountainous: According to *Spencer v. Spencer*, 24 Wash 2nd 574 (1946), UDSA representatives "had officially visited and examined the cave, declared that the department had been searching since 1925 for a locale having attributes of the caves in Roquefort, France, and that this cave was the first one of the desired kind that they had been able to find, and that this one, and no others in the United States, met the required qualifications" (577). Spencer quote: "Cheese Cave of Mt. Adams," Patsie Sinkey, *Seattle Daily Times*, September 24, 1944.

At this point the tale of Guler Cheese: *Spencer v. Spencer*, 24 Wash 2nd 574 (1946). See also "Motion Filed for Retrial of Spencer Case," *Goldendale Sentinel*, January 18, 1945; "Court News Given," *Goldendale Sentinel*, February 22, 1945.

After the divorce the Dean family withdrew: See Nancy Morris, "Created in Sunken Caves at Trout Lake: A New Cheese," *Oregonian*, November 18, 1951; *Seattle Daily Times*, June 21, 1953. Ads for Black Rock cheese appear in the *Seattle Daily Times* during 1952-53. Revival attempt: "Mt. Adams Cheese Plant May Rival Roquefort Production," *Oregonian*, May 11, 1956.

Near Menlo in the town of Lebam: Info on Dobler family from "Raymond is Young But Thriving City," *Morning Oregonian*, September 28, 1915; "Farm Follows Mill in Willapa Valley," *Morning Oregonian*, November 24, 1929.

In 1914 another group of businessmen: "Menlo," *Sou'Wester*, Pacific County Historical Society, XLII no. 2 & 3, (2006): 26.

In 1932 a group of investors approached: On the opening, see "Cheese Plant to Open this Week," *Central Point American*, February 2, 1933 and February 9, 1933. The plant is variously referred to as "Rogue River Cheese & Produce" and "Rogue River Cheese & Products"; ads appear throughout 1933 pleading for more milk. See, for example, *Central Point American*, April 13, 1933; May 25, 1933; June 9, 1933. "Central Point Cheese Factory Sold," *Medford News*, December 6, 1935. Ignazio "Ig" Vella historical narrative, n.d., collection of Rogue Creamery.

According to Ignazio "Ig" Vella: "Oregon Blue," *Eugene Register-Guard*, July 12, 1995; Ignazio Vella narrative.

In the dairy world, J. L. Kraft had already: Oregon Agricultural Experiment Station, *Canning Cheese*, Bulletin No. 78 (March 1904). "What Does Canned Cheese Offer the Industry?" *National Butter and Cheese Journal,* April 10, 1934, 6.

A number of large commercial manufacturers: For background on the era of canned cheese see *National Butter and Cheese Journal,* April 25, 1933 ("Canned Cheese Soon to Be Introduced by Grove City Creamery"); July 10, 1934 ("A&P Begins Canning of 200,000 Pounds American Cheese"); March 25, 1935 ("Land o' Lakes Now Canning American Cheese"); December 25, 1935 ("Canned Cheese Being Sold by Portland, Ore. Creamery"); "Name Chosen for Cheese Sold in Cans," *Morning Oregonian,* October 12, 1935; Loren H. Millman, "Oregon Cheese Put Up in Cans to Be Sold in Eastern Market," *Morning Oregonian,* July 31, 1938. In 1940, the Whatcom County Cooperative Dairy Association plant in Lynden, Washington, also produced and shipped fourteen million tons of canned cheese.

Meanwhile, at Washington State College in Pullman: Some researchers now believe that improved sanitation practices, rather than the adjunct culture per se, enabled Golding and his team to perfect the cheese canning process. See U. S. Patent No. 2,227,748 (filed April 19, 1939) (Golding's patent for canned blue cheese). In 2011 Washington State University sold $5.5 million worth of cheese, mostly Cougar Gold but several other styles as well, packaged in its signature metal cans. See Becky Kramer, "WSU Shipping Tons of Cougar Gold Cheese," *Seattle Times,* December 20, 2011.

As during World War I: *Seattle Daily Times,* July 16, 1942 (360 million pounds cheese sent to Europe); USDA, Total US Annual Cheese Production 1919-2010.

According to manager Carl Haberlach: "Tillamook Cheese Sold," *Oregonian,* October 29, 1941 (1941 total). "Cheese Ouptut Shows Increase," *Oregonian,* June 28, 1942; "Milk Products at High Level," *Oregonian,* February 19, 1943 (1942 record total); Idaho Department of Agriculture, *Biennial Report* (1946-47).

During this period the federal government: *Seattle Daily Times,* January 20, 1944.

The Lend-Lease program siphoned off: "Soft Cheeses Due on Ration List," *Sunday Oregonian,* May 23, 1943.

Because meat was more heavily restricted: See e.g., "Recipes Given for Cottage Cheese," *Oregonian,* May 27, 1943.

As the war drew to a close: Retailer quote: "Cheese Ration Recess Asked," *Sunday Oregonian,* June 17, 1945. Wholesaler: "Cheese Supply Piles Up," *Seattle Daily Times,* July 12, 1945.

Goat cheese production gained some momentum: See menu for Hotel Pierre, 1945, collection of New York Public Library, accessible at: http://menus.nypl.org/menu_pages/57464. Itasca Goat Dairy is mentioned in *Dairy Goat Journal* February 1942. An ad for Ozark Capri Cheese appears in *The New Yorker,* November 15, 1947.

The Cheese Factory at Tenas Illihee: The island's name is derived from the native Chinook terms *tenas*, meaning small, and *illahee*, meaning earth or land. The name has been spelled many different ways over the years; during the early twentieth century the island was referred to both as Tenas Illahee and Tenas Illihee, but the name is currently spelled Tenasillahe (one word).

Willard N. Jones was a civil engineer by training: Author interview, Jessie N. Jones (granddaughter), October 2011; background on the Oregon Land fraud cases from "Willard N. Jones Pardoned by Taft," *Morning Oregonian,* June 4, 1912; Stephen A. Douglas Puter, *Looters of the Public Domain* (Portland: Portland Printing House, 1908).

Jones went on to operate a dredging company: See "Diking an Island," *Morning Oregonian,* February 1, 1908; "Homes and Farms Now Dot Lands Recently Reclaimed from Floods," *Morning Oregonian,* October 24, 1915. Velma Vlastelicia, "Clifton History," *The Alder Patch,* December-January 1977 (collection Jessie Jones). The *Alder Patch* was a hand-typed and -drawn historical newsletter produced by longtime Clinton resident Vlastelicia.

In 1920 Jones's son Robert: Info about Robert L. Jones from Jessie Jones, "The Women of Tenas Illihee," *Cumtux* 31 no. 1 (2011); see also "Beef Cattle Graze Nearby," *Morning Oregonian,* August 25, 1940; "Torrid Sun Adds to Flood Menace," *Morning Oregonian,* June 13, 1933.

A number of goat's milk cheesemaking operations: *Dairy Goat Journal* 19 no. 6 (1941): 6 (Breese of Estacada); American Dairy Goat Yearbook 1945, 15-16 (cooperatives in Portland and Brookings, Oregon). See also "Dairy Goat Breeders Hold Special Banquet," *Oregonian,* November 29, 1942 (brief mention of Brookings cooperative).

Perhaps the most ambitious goat cheese factory: "Goat Herds Dot Region," *Oregonian,* May 19, 1946.

Chapter 5

In November of 1949 a newly constructed: "Tillamook Cheese Industry to Dedicate New Headquarters," *Oregonian,* October 30, 1949; Paul H. Mandt, "Tillamook Comes into its Own," *Butter Cheese and Milk Products Journal,* July 1950, 28; "Leader Dies at Tillamook," *Oregonian,* October 29, 1949 (Haberlach obituary).

The sparkling new building marked: Later litigation revealed that the Tillamook County Creamery Association actually owned several discrete portions of the facilities, including a storage building.

Perhaps it was residual patriotic fervor: "Cheese Factory Largest in West," *Twin Falls Times-News,* March 1, 1945. "Way out West in Idaho," *National Butter and Cheese Journal* (July 1945): 32-34.

In a 1947 article, the *Seattle Daily Times*: "Battle Ground Plant Biggest Single Unit on Coast," *Seattle Daily Times,* October 19, 1947. *Seattle Daily Times,* May 6, 1928 (1928 numbers); *Morning Oregonian,* July 1, 1928 (night shift added); *Morning Oregonian,* January 5, 1931 (1930 output numbers). 1948 Annual Report, Tillamook County Creamery Association, collection Tillamook County Historical Museum (lists creamery output annually going back to 1914). The site of the former Battle Ground cheese factory now houses Anderson Dairy, an independent fluid milk processor.

Meanwhile, in 1945, the *Daily Chronicle*: "To Erect New Chehalis Plant," *The Daily Chronicle* (Centralia), November 29, 1945; "Cheese Plant to Be Opened," *Centralia Daily Chronicle,* August 22, 1940.

Not to be outdone, Lynden: "Cheese-Making Champs a Flop at Selling it," *Seattle Daily Times,* December 14, 1947; see also "Lynden Honors Milk Drinking French Premier," *Seattle Daily Times,* December 7, 1954 (calling the Lynden plant the largest in the world). "Bellingham Bay: Remarkable Showing for the Year 1901 Made in Whatcom County," *The Seattle Daily Times,* February 9, 1902; "Lynden in the Heart of Very Fertile Region," *The Seattle Daily Times,* March 31, 1912. The Bellingham plant, housed in the former Bellingham Brewery building, closed in 1967.

The rapid expansion of a number of food industries: "Cincinnati's Embalmed Meat," *New York Times,* July 25, 1899.

The otherwise pastoral world of dairy products: "Honesty in Trade," *Morning Oregonian,* June 17, 1897 (filled cheese annual numbers); "Oil is Found in Cheese," *Morning Oregonian,* April 21, 1909 (Newberg plant).

While the Pure Food and Drug Act: See, e.g., *U.S. v. Ten Cases Bred Spred,* 49 F.2d 87 (8th Cir 1931); *U.S. v. Forty Barrels and Twenty Kegs of Coca-Cola,* 241 U.S. 265 (1916).

Meanwhile, scientists were already devoting: For an early survey of nineteenth and early twentieth century research into pathogens in cheese, see F. W. Fabian, Ph.D., "Cheese as the Cause of Epidemics," *Journal of Milk Technology* 9 no. 3 (1946). During 1936, the trade publication *National Butter and Cheese Journal* published a series of articles in successive issues in which scientists from across the United States and Europe debated the question, "Should Milk be Pasteurized for Cheesemaking?" The articles provide a unique view of the prevailing winds of opinion about the issue during this period.

The trend toward regulation of cheese: Statistics: F. W. Fabian, Ph.D., "Cheese as the Cause of Epidemics," 134. See, for example, "Should Milk be Pasteurized for Cheesemaking?" *National Butter and Cheese Journal,* April 10, 1936, 14 (according to Prof. H. A. Bendixen of Washington State College, "Much of the cheese in the state of Washington at the present time is made from pasteurized milk") and May 10, 1936, 10 (according to Prof. W. H. E. Reid of University of Missouri, 95 percent of the cheese made in Missouri is made from pasteurized milk).

FDA action was inevitable not only: Prior to 1949, the FDA had created just three standard definitions for cheese; the rules for cheddar, Colby, and washed curd cheeses did not require a sixty-day aging period. For the current version of FDA rules regarding cheese, a number of which have been revised since 1949, see 21 CFR Part 133 "Cheese and Related Cheese Products."

We have the Goodyear Tire & Rubber Company: *National Butter and Cheese Journal* (May 1946): 40. Studies later showed that a substantial number of workers in the Pliofilm manufacturing plant in Ohio developed benzene-induced leukemia. Study of the so-called "Pliofilm cohort" contributed to the virtual elimination of benzene from industrial manufacturing and commercial applications. In an interesting 1938 article written before the Pliofilm era titled "Post-War Concentration in the Cheese Industry" (*Journal of Political Economy*, Vol. 47 No 6, 1938 823-45), William Nichols argues, among other things, that the packaging revolution in the cheese industry began with the popularization of processed cheese. Processed cheese was packaged in neat, tidy tin cans, a significant departure from what he calls "the hopelessly inefficient marketing of bulk cheese."

Pliofilm was a breakthrough product: John Denny, "New Cheese Type Favored," *Oregonian,* October 5, 1947. Another experimental product of the period, Pliowax, consisted of Pliofilm combined with wax, creating a coating for cheese that was said to have been stronger and more durable than wax. Ads: *Oregonian,* August 8, 1947 ("Menus Brightened"); *Seattle Daily Times,* September 11, 1947 ("Delights Housewives"); see also *Clatskanie Chief,* February 20, 1948 ("Ask Your Grocer for the New Pliofilm Wrapped Darigold cheese in One Pound Packages").

In the 1930s researchers at the University of Wisconsin: *National Butter and Cheese Journal* (June 10, 1934): 26. "News of Food," *New York Times,* January 21, 1948.

Mt. Angel, a small town in Oregon's: Creamery account books in the collection of the Mt. Angel Historical Society show the variety of products produced by the cooperative over the years. See also "Mt. Angel Plant Scores," *Morning Oregonian,* November 4, 1930; joint ad for Mt. Angel Cooperative Creamery and Safeway stores, *Morning Oregonian,* May 18, 1932 (Safeway deal); "Business Churns in a Big Way at McMinnville Creamery," *Oregonian,* November 14, 1977 (Darigold's use of brand).

Some of the era's cheesemakers found success: Background info from *National Butter and Cheese Journal* (September 10, 1933): 21, and "Special Cheese Profitable, Firm Finds" (October 10, 1933): 42. Retail facility: A. J. MacPherson, "Cheese, the Way You Like It," *Butter, Cheese and Milk Products Journal* (March 1951): 28. In 1939, the Hansen firm moved, or perhaps attempted an expansion, into Chehalis for a period; see "Cheese Firms are Combined," *Centralia Daily Chronicle,* December 20, 1939. 1950s: See advertisement for Frederick and Nelson, *Seattle Daily Times,* May 18, 1958, that includes a display of seven varieties of local cheese, including Danish Banquet Cheese.

Carl Hopperdietzel founded an eponymous cheese: "Factory Operator Picks Cheese Over Lumbering," *Post Register* (Idaho Falls), January 4, 1963; "Fremont to Get Cheese Factory," *Post Register* (Idaho Falls), January 18, 1953 (Hopperdietzel plans to open factory). St. Anthony was one of the many Idaho cities J. L. Kraft had visited on his journey through the state during the summer of 1922.

Farther north in Sandpoint, Idaho: Susan Toft, "North Idaho Is a Tasty Place," *North Idaho Sunday* (Coeur d'Alene), January 20, 1985.

The southern Oregon coast has a long history: Ad copy from *Van Nuys News,* October 13, 1955 (several misspellings corrected). Ads for Reedsport cheese (sold by Vons Grocery stores) appeared in a variety of publications across southern California, including the *Los Angeles Times*, from 1950 through 1970.

Nearby in Bandon, the Coquille Valley: Julie Tripp, "Bandon Abandons Cheese Business," *Oregonian*, February 12, 1987. "Eugene May Lose Cheesemaker to Bandon," *Eugene Register-Guard*, August 1, 1987. See also "Bid to Buy Oregon Cheese Factory Collapses," *Spokane Spokesman-Review*, June 24, 1987; "Tillamook Cheese May Buy Factory," *Eugene Register-Guard* May 5, 1987.

In 1988, the Bandon factory was resurrected: "Farmers, Tourists Smile Again in Bandon," *Eugene Register-Guard* August 16, 1988 (Bandon restarts). In 2012, developers announced that a new cheese factory was to be built in Bandon, resurrecting the long tradition of cheesemaking in the region. Face Rock Creamery (named after a local rock formation) is set to open in 2013, with Brad Sinko, son of former owner Joe Sinko, as head cheesemaker.

Goat's milk cheese production continued: Ads for Bearga: *Seattle Daily Times*, March 2-5, 1959; goat's milk, *Seattle Daily Times*, June –July 1961. Mystic Lake: Susan Schwartz, "Redmond Goat Dairy Began With Modern-Day Heidi's Little Pet," *Seattle Times*, November 23, 1969; "What's New," *Seattle Times*, February 16, 1970 (yogurt debuts); "Goats are Beautiful People," *Seattle Times*, September 5, 1971.

M. P. Eggers once referred to himself: Eggers information from "Chehalis Cheese Plant Truly Unique," *Chehalis Daily Chronicle*, November 15, 1958; "Cheese Plant is Swept by Chehalis Fire," *The Daily Chronicle*, (Centralia-Chehalis), April 7, 1956; "Briar Hills Dairies in Chehalis," *The Sunday Olympian*, July 30, 1978; Bill Moomau, author interview, November 2011.

Swiss immigrants brought their cheesemaking knowledge: Others include the J. B. Migg factory in Kennydale, Washington (now part of Renton), mentioned briefly in *Washington: A Guide to the Evergreen State* (Portland: Metropolitan Press, 1941), 323; The Portland Cheese Company of Portland, Oregon, run by Swiss emigrant Otto Friedli; see also Jo Evalin Lundy, *Tidewater Valley: A Story of the Swiss in Oregon,* a young adult novel involving a Swiss family that moves to the Tillamook, Oregon, area in hopes of starting a cheese factory.

Still, swiss cheese production lagged: Bureau of Animal Industry, *24th Annual Report for the Year 1907* (Washington DC: Government Printing Office, 1909), 74. Stephanie Clark, et al., *The Sensory Evaluation of Dairy Products* (New York: Springer Science + Business Media, 2008), 431. *Butter, Cheese and Egg Journal* (June 29, 1921): 43.

Brothers Ernest, Paul, and Fred: Brog family background from Paul Schwartz, "The Brog Brothers and Their Switzerland in the Rockies," *National Butter and Cheese Journal* (February 1941): 12; "Salmon Swiss Cheese Factory Expands to Candy Operation," *Post Register* (Idaho Falls), February 27, 1968; "Purchases Interest," *Post-Register* (Idaho Falls), December 5, 1971; "Idaho Cheese Producer Plans to Develop Greater Variety," *Idaho Sunday Journal* (Pocatello), December 4, 1966.

The Brogs helped establish a regional pocket: "Swiss Cheese Factory at Wayan to be Big Help to Many Farmers," *Soda Springs Sun*, June 1, 1933 (Gray's Lake plant); Idaho Department of Agriculture 9th Biennial Report 1935-36 (area factories producing one-fourth of total U.S. production). Other factories producing swiss cheese in the area include the Teton Valley Swiss Cheese Company in Tetonia and the Nelson-Ricks plant in Driggs.

After World War II the Brogs consolidated: "Cheese Business Full of Holes," *News-Dispatch* (Jeannette, Pennsylvania), May 20, 1954.

In 1973, Idaho cheese entrepreneur Paul Brog Jr: "Age Old Art Practiced at New Cheese Factory," *Idaho Free Press*, March 31, 1973. Ed Gossner Sr. was once manager of the Cache Valley Dairy Association but later split off and founded his own operation during the 1960s; Don Shaff, "Cheese Production Booms at Nampa, Caldwell Plants," *Idaho Statesman*, March 29, 1976.

In 1990, the J. R. Simplot Company: *Idaho Press-Tribune*, April 25, 1993.

Meanwhile, scientists were hard at work: See, e.g., University of Wisconsin Experiment Station, *A Swiss Cheese Trouble Caused by Gas Forming Yeast*, Bulletin No. 128, (1905); United States

Dept. of Agriculture, *The Use of Basillus Bulgaricus in Starters for Making Swiss and Emmenthal Cheese*, Bulletin No. 148 (1915); U. S. Patent 2,494,636 (filed June 15, 1946)—patent for rindless swiss cheese process granted to Kraft, 1950. Small plants close: See Vivian Simmons and Ruth Varley, *'Gems of Our Valley,' A Written and Pictorial History of Gem Valley, Idaho* (Grace Literary Club 1977), 115-17 (Grace Lake); "American Cheese Now Being Made at Gray's Lake," *Soda Springs Sun*, October 15, 1953; "Grays Lake Cheese Factory Closes Down," *Caribou County Sun*, April 23, 1959. "Star Valley Cheese Shuts Down," *Billings Gazette*, November 30, 2005. "Purchases Interest" (Brog history at Salmon Valley plant), *Post-Register* (Idaho Falls), December 5, 1971.

In 2010, United States manufacturers produced: USDA - National Agricultural Statistics Service, Total U.S. Annual Cheese Production 1919-2010,.

For years, small markets in these ethnic: See, for example, John Mariani, *How Italian Food Conquered the World* (New York: Palgrave Macmillan, 2011).

During the first decade of the twentieth century: "Imperial Cheese to be Made Locally," *Chehalis Bee-Nugget*, July 17, 1936; "Cheese Company Has New Owners," *Centralia Daily Chronicle*, October 2, 1936 (mentioning acquisition of Morton plant).

Though the Depression took its toll: The Werthmanns, a husband and wife team, continued operating the Seattle Cheese Company for several years; see Dorothy Neighbors, "Seattle Factory Makes Rich, Piquant Cheeses," *Seattle Daily Times*, December 10, 1948.

Pat Castrilli was not out of the cheesemaking business: Information on Castrilli operations from: Joy Castrilli, author interview, April 2012; "Cheese Firm Moves into New Plant," *Seattle Sunday Times*, Dec 15, 1963; "Italian Cheese By Tons," *Seattle Times*, February 12, 1967; "They Have a Whey," *Seattle Times*, October 26, 1977.

Charles' son Louis took over in 1934: See "Orting Cheese Plant Saves Small Dairies," *Sunday News Tribune* (Tacoma), November 25, 1956; "Orting Factory Famous for its Three Cheeses," *Seattle Sunday Times*, January 27, 1963.

In 1989, CEO Darlyne Mazza, who took over: "Washington's Oldest Cheese Factory to Triple Production," *Moscow-Pullman Daily News*, October 8, 1987 (Mazza plant moving and expanding); "Sumner Cheese Plant to Close," *Tacoma News Tribune*, November 6, 1998; Lorianne Denne, "Cheese Factory Changes Hands as Finances Curdle," *Puget Sound Business Journal*, January 28, 1991. The Scardillo family, owners of Scardillo Cheese Company in Burnaby, British Columbia, also operated a cheese factory specializing in Italian style cheeses in Tacoma, Washington, for several years during the 1980s.

The seeds of the dispute had been sown: Beale Dixon, oral history OH 87.4, collection of Tillamook County Pioneer Museum.

In October of 1962, just a day before: "Controversy of Dairy Group Hits Courts Here," *Tillamook Headlight-Herald*, October 14, 1962 (lawsuit filed); "Judge's Decree Closes Lawsuit Between Coops," *Tillamook Headlight-Herald*, September 22, 1963 (entire judge's opinion printed in the paper).

Litigation flew back and forth for years: *Tillamook Headlight-Herald*, August 16, 1964 (TCDA files lawsuit); February 21, 1965 (TCCA files counterclaim). "Cheese Makers' Tiff Poses Store Problem," *Oregonian*, June 28 1964; *Tillamook County Creamery Association vs. Tillamook Cheese and Dairy Association* 345 F.2nd 358 (9th Cir 1965).

During the 1950s, prices for fluid milk: Beale Dixon, oral history, Collection Tillamook County Pioneer Museum. Floyd Bodyfelt, author interview, March 2012.

The dispute grew to epic proportions: "Reward Offered in Threat Case," *Tillamook Headlight-Herald*, May 12, 1963. "Bomb Blasts Home, Auto of Tillamook Dairy Chief," *Oregonian*, May 20, 1960. Archie Satterfield, *The Tillamook Way: A History of the Tillamook County Creamery Association* (Tillamook: Tillamook County Creamery Association, 2000): 91. Ad from *Tillamook Headlight-Herald*, March 22, 1964.

After severing ties with the TCCA: "Slowdown Halts Tillamook Plant," *Spokane Spokesman Review*, March 31, 1965. "Cheese Suit Continued," *Oregonian*, February 15, 1964.

TCDA eventually entered into a contract: "Foremost Dairy Inks Contract to Age, Market TCDA Cheese," *Tillamook Headlight-Herald,* September 5, 1965; "TC&DA to Do Own Cheese Marketing," *Tillamook Headlight-Herald,* April 2, 1967. "TC&DA to Triple Its Cheese Plant Output," *Tillamook Headlight-Herald,* April 27, 1967.

Swiss native Adolph Woodrich lived in Idaho: "Cheese Plant Ready Soon," *Morning Oregonian,* September 25, 1930. Additional information on Woodrich from oral histories of: Dick Chamberlain, Jessie Tyser, Ed Dahack, in "Eagle Point, Jackson County, Oregon Inventory of Historic Properties, 1990, Cheese Factory, 336 N Royal," collection Southern Oregon Historical Society.

Under investigation by the FBI: The story of Arthur Bell, Mankind United, and Christ's Church of the Golden Rule was covered extensively in the local and national press. See, for example, "Church to Start Cheese Factory," *Oregonian,* January 15, 1945; "Receiver Sought on Church Goods," *Oregonian* November 24, 1945; "Big Church Empire Bared," *Oregonian,* November 25, 1945; "New Religious Group Buys Up Vast Holdings," *Los Angeles Times,* March 2, 1944; "Profit's Prophet," *Time,* May 21, 1945.

There were several unexpected twists: "State Orders Recall of Contaminated Cheese," *Oregonian,* June 30, 1967; "Toxic Cheese is Withdrawn," *Seattle Times,* June 30, 1967; "Lab Absolves Suspect Cheese," *Oregonian* August 18, 1967. According to the *Seattle Times,* local health authorities seized a number of lots of TCCA cheese, many of which were found to be contaminated. Apparently later testing, as reported by the *Oregonian,* proved otherwise. Sanitation: The Sanitary Authority ordered a waste treatment plan, but not surprisingly, the TCCA and TCDA were unable to agree on one. In December of 1967, the *Oregonian* reported that TCDA had been denied a waste permit, while the TCCA was allowed one based on an abatement plan it had developed. See "Smog, Sewage Top Chronic List Tackled by Sanitary Authority," *Oregonian,* April 26, 1967; "New Antipollution Effort Aimed At Clearing State's Water, Air," *Oregonian,* December 29, 1967.

In September of 1968, the TCCA and TCDA: "Creameries End Dispute Over Tillamook Cheese," *Oregonian,* September 25, 1968; "Creameries Reach Agreement," *Tillamook Headlight-Herald,* September 26, 1968.

Chapter 6

During the 1960s and '70s communities: Plowboy Interview, John Shuttleworth, *Mother Earth News* #2, March 1970, 6.

It was not much of a stretch to connect: Steve Ellis, "Organic Farm Is Teacher's Classroom," *Seattle Times,* June 8, 1975; see also Jeffrey C. Sanders, "The Food Movement's Role in Revitalizing Environmentalism," *Seattle Times,* March 4, 2011.

The so-called "back to the land" movement: Roy Reed, "Rural Areas Population Gains Now Outpacing Urban Regions," *New York Times,* May 18, 1975; Kenneth M. Johnson, "Rural Rebound," *Report on America* (Population Reference Bureau) 1, no. 3, August 1999. The long history of the back to the land movement in the United States is covered extensively in Dona Brown, *Back to the Land: The Enduring Dream of Self Sufficiency in Modern America* (Madison: University of Wisconsin Press 2011). It should be noted that the back to the land movement was but one part of a complex socioeconomic and cultural trend pattern responsible for the population shift.

Few (if any) of this new generation: "Get a Goat!" *Mother Earth News* 6 (1970), 19.

Rural Polk County, west of Salem: Information on Singing Winds Goat Dairy from Peggy Sand, "Goat Herd Grows Quickly," *Oregonian,* November 4, 1977; Patrick Bleck, "A Goat for All," *Oregonian,* January 1, 1984; Bethanye McNichol, "1,200 Cloven Hooves Keep Things Astir at Singing Winds Goat Dairy," *Oregonian,* July 9, 1985; "Smile Kid and Say 'Cheese,' " *Eugene Register-Guard,* June 26, 1985.

In 1979 Sally and Roger Jackson received: "Small Energy Saving Projects Given Grants," *Seattle Times,* December 14, 1979; additional information on Sally Jackson from author interview,

April 2008. Nancy Harmon Jenkins "As American as Apple Pie," *New York Times,* March 1, 1992.

In rural Graham, Washington, east of Tacoma: Background on Kapowsin Dairies from "Goat Farm Becomes Cheese Dairy," *The Dispatch* (Eatonville, Washington), June 3, 1981; "Cheese: Couple's Dairy Adds Choice Offerings," *Seattle Times,* March 7, 1982. Elizabeth Rhodes, "The Market Harvest Never Ends," *Seattle Times,* November 5, 1980 (Greatorex quote); Marian Burros "Food Notes," *New York Times,* May 19, 1982. Several other goat's milk cheesemakers in Washington came and went during this period, including Marbelmount Goat Dairy in the small Skagit Valley town of Rockport and Feather Haven Dairy in Port Angeles on the Olympic Peninsula.

True to the spirit of the era: See "Goat Dairy Co-op Urged," *Sunday Oregonian,* November 8, 1970; Bill Lynch, "Goatmen Get Good News," *Eugene Register-Guard,* March 16, 1971; "Student Runs Goat Dairy Co-op," *Sunday Oregonian,* August 29, 1971; B. J. Noles, "Goat Cheese Co-op Gives Needy Help," *Sunday Oregonian,* May 4, 1972; Ann Baker, "Four Self Help Projects Chosen for Funding," *Eugene-Register Guard,* July 14, 1972.

While many of the Pacific Northwest's emerging: Background on Pleasant Valley Dairy from Joyce Snook (daughter of George Train), author interview July 2008; "Dairy Farm Survives," *Seattle Times,* April 10, 1980; Stuart Eskenazi, "Cheese Lover in Paradise," *Seattle Times,* May 12, 2005.

Those who live in the Pacific Northwest: Info on Tanzer from Barbara Durbin, "Portland Deli Owner Initially Had to Search Out Speciality item Suppliers," *Oregonian,* March 14, 1984, and author interview, Elaine Tanzer, July 2012; Bolson quote from Elizabeth Rhodes, "The Market Harvest Never Ends" *Seattle Times,* November 1, 1980. Loevenbruck: "Say Cheese: Importer Has French Contacts," *Oregonian,* June 12, 1977. Eurobest was purchased by DPI Group, a national distribution company, in 2000.

Consumer tastes were evolving: Bureau of Transportation Statistics, Air Carrier Traffic Statistics, accessed at http://www.bts.gov/programs/airline_information/air_carrier_traffic_statistics/ ("Revenue Passenger Emplanements"). "State Grape, Wine Industry Stable," *Seattle Times,* August 13, 1961 (8 wineries).

Given the growing popularity of all things European: Jim Kadera, "Velvety Brie Made in Old Barn New Oregon Enterprise," *Oregonian,* November 2, 1979; Mike Henderson, "Blue Heron Fromage de Brie," *Seattle Times,* August 3, 1980; "Potpourri," *Oregonian,* August 5, 1986 (award). Nehalem Bay Winery is located in the building that formerly housed Mohler Creamery, once part of the Tillamook Cooperative.

Pierre Kolisch left the legal profession: Greg Higgins, author interview, January 2009.

Four dairy farmers started Yakima Valley: "Cheese Factory To Be Built," *Tri-City Herald,* June 1, 1984; Rick Larson, "Gouda Going Great," *Tri-City Herald,* August 14, 1985.

Farmers markets did not disappear entirely: "Low Cost Market Opening," *Seattle Times,* July 30, 1976; Theresa Chebuhar, "Farmers Markets Are Blossoming," *Seattle Times,* July 8 1979. "Farm and City," *Oregonian,* July 23, 1983 (small market in north Portland opens); Jonathan Nicholas, "At Last, Stumptowners Get Spuds With Soil Attached," *Oregonian,* June 18, 1992 (downtown Portland farmers market); Washington, Oregon, and Idaho farmers market numbers from Washington State Farmers Market Association, Oregon Farmers Market Association, Idaho State Department of Agriculture. It should be noted that the USDA's Farmer-to-Consumer Direct Marketing Act of 1976, which provided money to states to promote direct farm to consumer marketing, was a major force in the growth and proliferation of farmers markets during this period.

Farmers markets have grown into: Rod Volbeda, author interview, May 2007. Sarah Marcus, author interview, March 2012.

People still talk about the James Cook: Theresa Simpson, author interview, January 2012.

The rapidly increasing supply and diversity: Steve Jones, author interview, April 2012.

During the nineteenth and early twentieth century: U. S. Department of Agriculture, National Agricultural Statistics Service, Total U.S. Annual Cheese Production 1919-2010.

Sartori: In March of 2013 Sartori announced that its Blackfoot Plant had been sold to Glanbia foods.

Sorrento Lactalis: Expansion news and facts from "Canyon County's Lactalis Cheese Plant Announces $40 million Expansion, 70 New Jobs," *Idaho Statesman*, February 6, 2012.

While industrial-scale cheese production: "Darigold Closes Dairy Fair Gift Shop, Cafe," *Yakima Herald-Republic*, February 24, 2012 (Darigold production numbers); "Tillamook CEO Reveals Strategy for Business," *Oregon Business*, January 2010 (Tillamook numbers). See "TCCA layoffs take Effect February 4th," *Tillamook Headlight-Herald*, January 27, 2012; Lori Tobias, "Much of Tillamook Cheese Factory's Packaging Operations Shut Down," *Oregonian*, February 4, 2012.

Despite the overwhelming dominance of industrial: See, e.g., Barbara Hansen, "Boutique Goat Cheeses from Idaho," *Los Angeles Times*, December 19, 1991. The Gunner-Pickens family also operated a small goat dairy called Goat Mountain Cheese Company for several years in Porthill, Idaho, near the Canadian border. The family closed their operation in 2005.

One reason that Idaho is home to relatively: Idaho Department of Agriculture, *10th Biennial Report 1937-1938*, 35. Stacie Ballard, author interview, February 2012.

Jeffrey Roberts estimates that in 1990: Jeffrey Roberts, *The Atlas of American Artisan Cheese* (White River Junction, Vermont: Chelsea Green Publishing Co. 2007), xx, and author interview, Jeffrey Roberts, June 2012; data on number of cheesemakers in Oregon, Washington, and Idaho past and present from: Washington State Department of Agriculture March 2012; Oregon Department of Agriculture March 2012; Idaho Department of Agriculture March 2012.

The artisan cheesemaking community: For more specifics on contemporary Pacific Northwest cheesemakers and the variety of cheeses they produce, see Sasha Davies, *The Guide to West Coast Cheese* (Portland, Oregon: Timber Press, 2010) and Tami Parr, *Artisan Cheese of the Pacific Northwest* (Woodstock, Vermont: Countryman Press, 2009).

In 2010 total United States cheese production: On growing U.S. consumption of mozzarella cheese, see Michael Moss, "While Warning about Fat, US Pushes Cheese Sales," *New York Times*, November 6, 2010; on production of artisan cheese in the United States, see Heather Paxon, "Locating Value in Artisan Cheese: Reverse Engineering Terroir for New-World Landscapes," *American Anthropologist* 113 (2010): 454 (Paxon estimates total United States artisan cheese production in 2006 at eight to ten million pounds). Current production: Jeff Roberts, author interview, June 2012.

In 1918 area dairy farmers organized: "Toledo to Celebrate," *Morning Oregonian*, May 11, 1919; "Cowlitz Cheese Plant to Open" *Morning Oregonian*, June 6, 1919.

Arthur Karlen, owner of Valley Creamery of Oakville: "Karlen Creamery at Toledo Handles Much Milk," *Chehalis Bee-Nugget*, June 10, 1932 (describes the various properties owned by Karlen).

The town of Toledo nevertheless: "Cheese Day Goal to Entertain Family," *Daily Chronicle* (Centralia-Chehalis), July 17, 1977.

Despite the relatively small scale: Paul Kindstedt, *Cheese and Culture* (White River Junction, Vermont: Chelsea Green Publishers, 2012), 208-9. William Neumann, "Raw Milk Cheesemakers Fret Over Possible New Rules," *New York Times*, February 4, 2011.

Raw milk cheese has developed into a particularly: Much was written in the local and national press about these incidents. See, for example, "Small Cheesemaker Defies FDA Over Recall," *New York Times*, November 20, 2010 (Estrella); Lynn Terry, "Portland Food Detectives Crack E. Coli Mystery," *Oregonian*, December 23, 2010 (Jackson). In November of 2012 a federal court issued a permanent injunction barring Estrella Family Creamery from selling cheese outside the state and ordered them to pay the costs of the 2010 seizure.

The state of Idaho has taken a unique approach: For an in-depth treatment of the raw milk issue in Idaho, see Guy Hand, "The Raw Milk Deal," *Boise Weekly*, November 30, 2011.

Among the other significant issues: Gianaclis Caldwell, author interview, February 2012.

For those in business already, the challenge: Carine Goldin, author interview, March 2012.

Another ongoing challenge for the region: Social media funding is a new trend in the artisan cheese world; to date only a few have taken advantage of it to fund a cheese business in some manner, though the numbers are on the rise. Belle Chevre of Alabama sought to raise $100,000 toward the purchase of a plot of land in late 2011, but was unsuccessful. In early 2012, Hawaii's Naked Cow Dairy raised over $18,000 toward the establishment of a cheese facility at their already existing cow dairy.

Even as the Pacific Northwest's artisan cheese: Laura Werlin, author interview, March 2012.

The key to a prosperous future: Jillian Greenawalt, author interview, October 2012. Sluder started her cheesemaking operation by trucking milk to Steve and Stacie Ballard's nearby farm, and the Ballards made her cheese for several years before Sluder developed her own on-farm facility. Full Circle Creamery: Kate Humiston, author interview, October 2012. In early 2013 Green Goat Dairy merged with Blue Sage Dairy, and Blue Sage Dairy currently produces both sheep and goat's milk cheeses.

Perhaps what is most remarkable about: Sherwin Ferguson, author interview, March 2012.

Appendix A

In 1799, reigning Tsar Paul I granted: James Gibson, *Imperial Russia in Frontier America: The Changing Geography of Supply of Russian America 1784-1867* (New York: Oxford University Press 1976), 94 (early settlement on Kodiak Island), 102 (number of livestock on Kodiak Island), 102-11 (discussion of the many problems the Russians encountered trying to farm in the region).

The ongoing task of maintaining a steady flow: *A History of the Russian American Company* A. Tikhmenev, trans. Richard Pierce and Alton S. Donnelly (Seattle and London: University of Washington Press 1978), 226. See also Gibson, *Imperial Russian in Frontier America,* 125-39 (detailed discussion of the various problems encountered in farming operations at Fort Ross).

Many who traveled to Alaska seeking gold: Comparing Alaska's agricultural potential with Finland: Frank G. Carpenter, "Southern End of Alaskan Railroad Opens 400,000 Acres of Rich Land," *El Paso Herald*, April 22, 1916. See also "Agriculture in Alaska," *Morning Oregonian*, December 6, 1900; "Real Estate," *The Seattle Daily Times,* January 20, 1900; Alaska: Hearings before the Committee on the Territories, House of Representatives, Sixty-Seventh Congress, First Session on HR 5694 (Washington: Government Printing Office, 1921). Ad for Valdez Dairy: *Valdez News*, December 7, 1901. Details about Hinckley from: "Personal Mention," *Fairbanks Sunday Times*, September 13, 1908; "Is Raising Stock in the Tanana Valley," *Fairbanks Daily Times*, September 20, 1908. Charles Creamer purchased Hinckley's Dairy in 1928 and Creamer's Dairy operated in Fairbanks through the 1960s. On the additional dairies in Fairbanks, see "Merry, Merry Milkmaid War," *Fairbanks Daily News-Miner* June 4, 1914. Details about the Juneau area's dairy history: Draft Report of the Juneau Dairy Farming Historic Resources Survey, City and Borough of Juneau, Alaska, Community Development Department, 1991. Simenof Island: "Alaska News Notes," *Fairbanks Daily News-Miner*, July 11, 1917.

Despite the efforts of Alaska's earliest dairy: Alaska: Hearings before the Committee on the Territories, House of Representatives, Sixty-Seventh Congress, First Session on HR 5694 (Washington: Government Printing Office, 1921). The type of cheese produced was not specified.

In an effort to encourage local agriculture: Background on experiment station history from *Agroborealis*, Vol. 30 No. 1 Spring 1998 (published by the Agricultural and Forestry Experiment Station, University of Alaska Fairbanks). Quote: Annual Report of Alaska Experiment Stations for 1906 (Washington D.C.: Government Printing Office 1907), 18; Detailed discussion of making cheese at Kenai Station from Annual Report of Alaska Experiment Stations for 1907 (Washington D.C.: Government Printing Office 1908) beginning on page 71.

Livestock research was later moved: 1907 animal count: Annual Report of Alaska Experiment Stations for 1907 (Washington D.C.: Government Printing Office 1908) page 59. Resumption of butter and cheese production: Annual Report of Alaska Experiment Stations for 1911 (Washington D.C.: Government Printing Office 1912); discussion of work at Kodiak Station begins on page 30; quote on page 32. Discussion of moving cattle: Annual Report of Alaska Experiment Stations for 1912 (Washington D.C.: Government Printing Office 1913) beginning at page 67.

The Matanuska Valley Farmers Cooperative: The cooperative was controversial from the beginning. Many settlers resented the imposition of the cooperative by the Alaska Rural Rehabilitation Corporation (AARC), the administrative body set up by the federal government to administer the colony's affairs. Some believed that the cooperative did not deliver sufficient returns and many conflicts arose among members. In 1940, the AARC turned over management of the co-op to the community. See Kirk H. Stone, *Alaskan Group Settlement: The Matanuska Valley Colony* (United States Dept. of the Interior, Bureau of Land Management, Washington DC, 1950), Chapter 9.

The farmstead cheesemaking wave spreading: Jeremy Hsieh, "Goat Farms Spur Growth in Alaskan Cheesemaking," *Seattle Times* April 10, 2010; author interview, Cherie Lowry, dairy sanitarian, State of Alaska, October 2012.

As of 2012 the only cheese producer: Rindi White, "Ribbon Cutting Opens Creamery," *Alaska Daily News*, May 31, 2008; Margaret Bauman, "Creamery Pre-Sells 6.5 tons of Cheese, More to Come," *Alaska Journal of Commerce*, April 27, 2008. Matanuska Creamery faced a significant challenge right at the outset when forty thousand pounds of its cheese tested positive for listeria and eventually had to be destroyed. The company blamed the problem on the raw milk it used to make the tainted cheese. Other problems: See, for example, Todd Disher, "Another $200K For Creamery," *Mat-Su Valley Frontiersman*, March 6, 2010 (creamery receives its third operating loan from the state despite lack of collateral); Ellen Lockyear, "Is Alaska's Dairy Industry Making a Comeback? Or on the Brink?" *Progressive Dairyman*, June 7, 2011; K. T. McKee, "Complaint Cries Foul on Matanuska Creamery," *Mat-Su Valley Frontiersman*, May 27, 2011 (detailing wrongful termination complaints filed against the creamery); "Agriculture in the Valley is Our History, Our Future," *Mat-Su Valley Frontiersman*, May 15, 2011 (suggesting that Matanuska Creamery is seeking a new, less expensive facility); Andrew Wellner, "Milk Runs Dry," *Mat-Su Valley Frontiersman*, October 11, 2011 (creamery suffering due to local milk shortage).

Index

Page numbers in bold type refer to illustrations.

Abbie and Oliver's, 144
aboriginal peoples
    See indigenous peoples
Acequia Dairy and Produce Company, 71
Adams, Cecilia, 38
advertising, 42, 61, 79, 88, 156, 167
Aerni, Joseph, 96
aging cheese, 46, 94, 97–98, 113, 114–15, 120, 166
agricultural colleges, 56–58, 150
    See also Oregon State; University of Idaho; Washington State
agriculture
    European style introduced to Pacific Northwest, 18, 27, 31, 163
    industrial, 56–57, 134, 140
    post-WWI slump in, 83
    sustainable, 134, 135, 150, 160
Aguilar, Benedita, 151
Alaska, 163–68
Albany, OR, 140
Alberta Creamery Association, 68
alfalfa, 70, 155
allergies, 148
Alpine Lakes Sheep Cheese, 170
alpine-style cheese, 118, 150
Alsea Acres Goat Dairy, 169
Alvord, Henry, 46
American cheese, 117
American Cheese Society competition, 139, 140, 148, 151, 152
American Milch Goat Record Association, 78

Americans, 29, 46, 104, 138
Ancient Heritage Dairy, 149, 169
animal feed, 70, 155
Ankeny, Henry, 46
Appel Farms, 157, 170
appellation d'origine contrôlée (AOC), 158–59
Applegate, Cynthia Ann, 44, 135
Applegate, Jesse, 43, 44
Applegate Trail, 44, 99
artisan cheesemaking, 135–44, 147–52, 154–60, 167
Astor, John Jacob, 15, 16
Astoria, OR, 15, 42, 105
Atlas of American Artisan Cheese, 149
Auburn Creamery Co., 66
awards, 46, 139, 148, 152
Ayrshire cattle, 47

Babcock, Stephen, 56
Babcock test, 56
Back Country Creamery, 170
back to the land movement, 133, 134–35
bacteria, 58, 77, 102, 113, 130, 154
Ballard, Steve and Stacie, 148, 149
Ballard Family Dairy and Cheese Company, 148, 171
bamm ost cheese, 92
Bandon Foods, 118
Baranov, Alexandr, 163
Battro, Mariano, 158
Bearga Goat Dairy, 119
Bear Lake Valley, 50, 72

Bear Lake Valley Cooperative, 121
Beaver Cheese Factory, **59**
Beecher's Handmade Cheese, **142,** 143, 152, 158, 170
Bell, Arthur, 129
Beu, Gary and Carla, 167
Beus, Kyle, 167
bicycling, 58
Birmingham, William, 49
Bishop, William, 41–42
Blackmore, William, 62
Black Sheep Creamery, 170
bloomy-rinded cheeses, 139, 152, 157, 160
blue cheese, 91, 93–101, 102, 104
Blue Heron French Cheese Co., 139
Blue Rose Dairy, 170
Blue Sage Farm, 148, 158, 171
Boardman, OR, 147
Bogart, John, 49
Boise, ID, 42, 74, 140, 149
Boise Farmers Market, 140
Borden Company, 100, 103, 107, 116, 131
Boswell, Benjamin, 44–45
branding, 61, 80, 98, 109, 115, 127, 128, 141, 158–59
Branson, Winnie and Jay, 91, 94
breweries, 76
Brewster Cheese Co., 145
Briar Hills Dairy, 91, **119**–20
Briar Rose Creamery, **141,** 157, 169
brick cheese, 92, 117
brie cheese, 103
British Empire, 16, 29
Brog brothers, 121
Brown, Alson "A. L.," 65
Brown, Vicky, 156
Bryant, Cary, 142
Bryon, Ed, 42
Buchwalter, Margie, 167
Burlington, WA, **51**
butter, 23, 38, 40, 45–46, 56, **64,** 65, 68, 112
buttermilk cheese, 65

caciovallo cheese, 125
Cada Dia Cheese Co., 169
Caldwell, Gianaclis, 155
California, 16, 21, 30, 35, 79, 113, 144
Camembert cheese, 93
canned cheese, 87–88, 101–2
Capendale, William, 25
Capitol City Farmer's Market, 140
Capper-Volstad Act, 84
Carpenter, Helen, 38
Carpenter, W. H., 70–71
Cascade Mountains, 39, 69, 159
Cascadia Creamery, 170
Cascadian Farm, 134
Castrilli, Louis and Elizabeth, 124, **125**
Castrilli, Pasquale (Pat), 124
Catholic Church, 28
Catterlin, E. W., 94
cattle
    breeding of, 56
    imported into Pacific Northwest, 21, 30, 165–66
    kept by pioneers and settlers, 27–28, 36–37, 39, 40, 43, 45
    ranching in Pacific Northwest, 19, 21–22, 30, 47, 71, 73
    and tuberculosis, 76–77
    See also specific breeds
caves, 91, 94, 97–98
Central Point, OR, 100–101
Champion, Joseph, 40
Chattaroy Cheese Co., 170
cheddar cheese
    canned, 101, 102
    mass produced, 51, 92–93, 94, 103, 120, 137, 147
    process of making, 60, 93, 130
    similar to cheese made by HBC, 24, 26
Cheese Bar, 143
cheese brogies, 122
Cheese Cellar, The, 143
cheese curds, **11,** 43, 113, 122
cheese factories

See factory cheesemaking
cheese festivals, 152, 153
cheese hoops, 43, 44
Cheese Louise, 144
cheese presses, 44, 47, 56
cheese production
    artisan, 152, 154
    national, 52, 83, 93, 103, 123, 152
    in Pacific Northwest, 58, 83, 154,
        165
cheese shops, 122, 138, 141–42,
    143–44, 152, 168
cheese vats, 47, 56, 153
Chefs on the Farm (Misterley and
    Jurgenson), 156
Cheney Cheese Factory, **66,** 67
Cheney Creamery, 67
cheshire-style cheese, 24
Childs, B.F., 74
Chimacum, WA, 42, 49
Chinook Indians, 15
Christensen, Fred, 61
chymosin, 22
clabbered milk, 76
Clatskanie, OR, 15
Clatsop Factory Cream Cheese, 47
Clover Leaf Factory, 72, 109
Coast Range, 43
Coeur d'Alene, ID, 29
Coeur d'Alene Indians, 29
collaborative cheesemaking, 158
"colonization cheddar," 9
colonization of Pacific Northwest,
    14–16, 32
    See also territorial claims to Pacific
        NW
Columbia River, 13, 15, 17, 18, 29,
    42, 105, 133, 164
Colwood Farm, 31
Commercial Creamery Co., 147
commercial dairying
    See dairy industry
Community Supported Agriculture
    (CSAs), 156
Consolidated Creamery Company, 69
Constance Cove Farm, 31

Conway Family Farms, 170
Cook, Captain James, 13, 14
Cook, James, 141–42
cooperatives, dairy, 49, 51, 59,
    60–64, 84, 107, 130
Coos Bay, OR, 61, 63, 64
Coos Bay Creamery, 62
Coos Bay Mutual Creamery, **63**
Coos-Curry Cheesemakers Associa-
    tion, 63, 85, 95
Coquille Valley Dairy Cooperative,
    118
cottage cheese, 38, 89, 104
Cougar Gold Cheese, **82,** 102, 143,
    150
Country Morning Farm, 170
coverings, 87, 101, 102, 115, 131
    See also canned cheese; Pliofilm;
        rind; wax
Cowlitz Farm (Cowlitz Grazing
    Farm), 20, 21, 23, 25, 31, 153
Cowlitz River, 20
Cowlitz Valley Cheese Association,
    153
Cozy Vale Creamery, 170
Craigflower Farm, 31
Cranberry Ridge Farm, 167
Cranston, Warren, 46
Cranston's Cheese, 46
cream cheese, 47, 89
cream separation, 23, 56
CSAs (Community Supported
    Agriculture), 156
cults, 129
curd mills, 56
curds, **11,** 43, 113, 122
Curds & Whey, 143
curing, 47, 113, 120, **141**
Curry County, OR, 61, 94

Dairy Division of the Bureau of
    Animal Industry, 93, 120
dairy equipment, 44, 45, 47, 56, 60,
    111, 137
Dairy Farmers of Oregon, 151
dairy farming, 19, 42, 70, 72–73

dairy industry
    in Pacific Northwest, 51–52, 57,
        80, 144, 166
    in United States, 76, 93, 103–4,
        113, 124
dairy science, 56–58, 93, 115, 120,
    122–23
dairy workers, 25, **95**
The Dalles, OR, 27, 39, 44
Dammeier, Kurt, 143
dams, 70, 72, 109, 133
Danish Banquet Cheese Co., 117
Darigold, 85, 110, 111, 115, 147, 167
Davis, Alonzo, 42
Davisco, 145
Dean, Wade, 97–98
DeCastilhos, Flavio, **143**
Dee Creek Farm, 170
Dercyx, Woody, 134
DeSmet, Father Pierre-Jean, 28–29
distribution, 68, 89, 107, 157
Dixon, Beale, 126, 127, 128, 130
domesticated animals
    See livestock
Donation Land Claim Act, 36
Dufur, Andrew Jackson, 42
Dungeness, WA, 42
dutch (pot) cheese, 38, 43, 165
Dutch-style cheeses, 139, 152
    See also edam; gouda; havarti

Eagle Point, OR, 129
edam cheese, 92, 117, 139
Edelweiss Creamery, 123
Eggers, Melvin "M. P.," 79, 91,
    119–20
Eggers, Peter, **119,** 120
Ellensburg Creamery Company, 68
El Michoacano, 151, 170
Emmenthaler (swiss) cheese, 51, 97,
    120–23, 148
Encaria Goat Farm, **78,** 79, 91
England, 16, 29
English cheeses, 24–25, 32, 42, 142
environmentalism, 133–34, 160
    See also sustainable farming

enzymes used in cheesemaking, 22,
    28, 56, 76, 157
Estrella Family Creamery, 154
European-style cheeses, 93–94,
    116–17, 120–25, 138
    See also specific types
Evans, Chuck and Karen, 147–48
exploration of Pacific Northwest,
    13–15

factory cheesemaking
    in Idaho, 88, 144–47
    post-WWII expansion of, 109–11,
        130–31
    problems with, 154
    shift to from farmstead cheese-
        making, 52, 58–73, 80
    smaller regional factories, 92,
        116–26
    See also Tillamook Cheese Factory
Fairaview Farm, 170
Fairbanks, AK, 165
Fairview Farm, 169
Fares, Lucinda Collins, **41**
farmer's cheese, 38, 43, 165
Farmers Cooperative Creamery, 116
Farmers Institute, 57
farmers markets, 74–75, 123, 140–41,
    156, 168
farming
    See agriculture
farmstead cheesemaking
    costs of, 155–56, 158
    in Idaho, 147–49
    larger operations, 65
    related to back to the land move-
        ment, 135–37
    by settlers in Pacific Northwest,
        40–45, 46–47, 58–59
    shift away from to factory cheese-
        making, 52, 131
    See also artisan cheesemaking
Farmstead Creamery Advisor, 155
FDA (Food and Drug Administra-
    tion), 111–14, 154, 155
Ferguson, Sherwin, 159–60
Fern's Edge, 169

festivals, 152, 153
feta cheese, 106
filled cheese, 112
firkins, 23, 44
Florence, OR, 44
Foland, Merriman, 58
Food and Drug Administration
    (FDA), 111–14, 154, 155
food safety, 93, 112–13, 127, 130,
    134, 154
food standardization, 61, 111–14, 154
Fort Astoria, 15–16, 19, 22
Fort Boise, 23
Fort Colville, 19, 21, 27, 29
Fort George, 16, 18, 21, 22
Fort Hall, 23, 39
Fort Langley, 19, 21
Fort Nez Perce, 21, 23, 27
Fort Nisqually, 20, 21–22, 23, 24, 31,
    65
Fort Okanagan, 23
Fort Ross, 164
Fort Vancouver
    farm and dairy operations of, 19,
        20–21, 22–24, 25, 164
    HBC relocation from to Fort
        Victoria, 30–31
    site of, 18, 75
    as supply site for settlers and
        missionaries, 27
Fort Victoria, 30, 31
49th parallel, 30, 55
Foster & Dobbs, 143
Fraga Farm, 169
French Canadians, 24, 27
French-style cheeses, 139
    See also bloomy-rinded cheeses
Frisia Dairy & Creamery, 170
Full Circle Creamery, 158, 169
fungus, 93, 98, 101, 114, 115
fur trading, 15, 17, 26, 27, 30, 31,
    163, 164

Gale, Joseph, 30
Galloway cattle, 165–66
Georgeson, Dr. Charles, 165

Gilliam, Martha, 40
gjetost cheese, 92
Glanbia Foods, 145, 148
Glendale Creamery, **161**
Glendale Shepherd, 149, 170
Glengarry Cheese Company, 137
gloucester cheese, 24, 25
goats, 13–14, 77–79, 104, 135, 136,
    159
goat's milk, 78–79, 90, 92, 104, 119,
    120, 135, 147–48
goat's milk cheese
    artisan, 135, 157
    consumption of during and after
        WWII, 104, 106, 118–19, 131
    early production of, 79, 90–92
    made by French Canadian HBC
        employees, 24
Goddik, Professor Lisbeth, 150
Gold Creek Dairy, 165
Golden Glen Creamery, 170
Goldin, Carine, 155
Goldin Artisan Goat Cheese, 155, 169
Golding, Dr. Norman S., **82,** 102
gold rushes, 35, 55, 164
Goodwin, John, 68
Goodyear Tire & Rubber Company,
    115
Gossner Foods, 145
Gothberg Farms, 157, 170
gouda cheese, 137, 139, 143, 152
Grace Harbor Farms, 170
Grande Ronde River, 37
Gray, Captain Robert, 13
Greatorex, David, 137
Greenawalt, Jillian, 158
Green Goat Dairy, 148, 158, 171
Green River Cheese Co., 124–25
Gremmels, David, 142
Griffon, Paolo, 62
grocery stores, 75, 116, 140
Groupe Lactalis, 144, 146
Grove City, PA, 93, 101
Growing a Farmer (Timmermeister),
    156
Grumpy's Goat Shack, 171

Guler Cheese Co., **97**–98

Haberlach, Carl, 59–60, 74, 103, 109
Hamilton Cheese Factory, **125**
Hanna, Henry, 47, 50
Hansen, Christian, 56
Hansen, Hans, 95–96
Hansen, John, 44
Harris, W. C., 62
Harry & David Corporation, 96, 118
Haun, Catherine, 37
havarti cheese, 139
Hawaii, 14
    See also Sandwich Islands
Hazelwood Creamery, 67–68
Hazelwood Cream Store, 67
Healing Acres Goat Dairy, 167
Heron Pond Farms, 170
Herron Hill Dairy, 170
Hinckley, Charles, 165
Hinckley's Dairy, 165
Hobson, John, 43
Holstein cattle, 47, 56, 93
Holstein cheese plant, 109
Homestead Act, 36, 70, 164
homesteaders, 31, 40, 45, 52, 105,
    135, 153
Hopperdietzel, Carl, 117
Hubbard, Willis, 67
Hudson's Bay Company (HBC),
    16–26, 29–32
    See also Puget Sound Agricultural
    Company
Hueffed, Stephen, 149
Humiston, Brian and Kate, 158
hydroelectric power, 83, 133

Idaho
    cheese production of, 45, 50, 71,
        88, 103, 144
    dairy regulations of, 154–55
    factory cheesemaking in, 71,
        144–49
    state of, 55, 70, 134, 164
    University of, 57, 67

Idaho Dairymen's Association, 72
Idaina cheese, 122, 123
Imperial Cheese Company, 124
indigenous peoples
    conflict with settlers, 26, 27, 39,
        40, 44
    before European colonization, 14,
        15, 163
    and missionaries, 28–29, 164
industrial cheese production
    See factory cheesemaking
irrigation, 69–70, 72, 88
Italian-style cheeses, 123–26, 146,
    152
    See also caciovallo; jack; mozza-
    rella; provolone

Jackson, Sally, 137, 154
jack-style cheese, 117, 124, 125, 130
Jacobs Creamery, 170
James Cook Cheese Shop, 141–42
Jerome Cheese Company, 145
Jerome Cooperative Creamery,
    109–10
Jersey cattle, 42, 47, 56, 93, 148
Jesuits, 28–29
Jones, Steve, 143, 144
Jones, Willard N., 105
Juniper Grove Farm, **132,** 139, 169
Jurgensen, Karen, 156

Kaeser, John, 72
Kahn, Gene, 134
Kapowsin Dairies, 137, 138
Karlen, Arthur, 153
Katmai, Mt., 166
Keister, William, 44
Kelley, Hall, 35
Kent, F. L., 57
Kent, Pete, 151
King County, WA, 41, 66
Kittitas Creamery Company, 68
Kittitas Valley, 44, 68
Klamath River, 70
Koch, Henry, 47–49
Koch, Robert, 77

Kodiak Island, 163, 166
Kolisch, Pierre, **132,** 139
Kraft, James L., 87–88, 101
Kraft Foods, Inc., 87–90, 100, 103,
    110, 117, 123, 145, 167
Kunz Brothers Dairy, 72
Kunze, Gustav, 72
Kunz family, 72–73
Kurtwood Farms, 156, 170

Ladino Cheese Factory, 129
La Grande, OR, 45
La Mancha goats, 159
La Mariposa, 158, 169
land-grant schools, 56–58, 150
    See also Oregon State; Washington
        State; University of Idaho
land stewardship, 133–34, 160
    See also sustainable farming
Lane County, OR, 44
Langlois, OR, 94–96, 98
Langlois Blue Star Cheese, 96
Larkhaven Farm, 170
Lark's Meadow Farm, 148, 171
Laughlin, Mary Yeargin, 39
Laura Werlin's Cheese Essentials, 158
Lee, Jason, 26–27, 30
Lett, David, 138
Lewis, Meriwether, and Clark, Wil-
    liam, 26, 35
Lewis County Dairymen's Associa-
    tion, 99, 110
Lewis Pacific Dairymen's Association,
    110
Lewiston, ID, 27
Litehouse Foods, 118, 145
Little Bear Dairy, 155
Little Brown Farm, 156, 170
livestock
    feral, 14, 16, 30
    importation of to PacificWNorth-
        west, 14, 16, 21, 22, 30, 36,
        163, 165
    as part of oceangoing expeditions,
        13–14, 21, 77
    See also cattle, goats

L. J. Ranch, 170
locavore movement, 138, 150
Long, J. Henry, 49
Louisiana Purchase, 26, 35
lumber, 40, 44, 63, 75, 79, 90, 99,
    124
Luthy, Fred, 48

MacKinnon, J.B., 150
Mama Terra Microcreamery, 150, 169
Manifest Destiny, 26, 30, 35–36
Manwaring Cheese Co., 171
Maple Leaf Creamery, 59, 109
Marcus, Sarah, **141**
marketing and sales
    mail order sales, 91, 96, 118, 119–20
    marketing cooperatives, 60–61, 85,
        109–10, 127, 128
    wholesale trade, 59, 67, 73–75
        See also advertising; branding;
            distribution
markets
    access and transport to, 43–45, 56,
        59, 64, 80, 149, 156, 168
    urban, 47, 49, 73, 139, 142, 149,
        168
        See also farmer's markets; grocery
            stores; retail stores
Martin, Francis M., 67
mass production
    See factory cheesemaking
Matanuska Creamery, 167
Matanuska Maid, 166–67
Matanuska Valley, 166–67
Matanuska Valley Farmers Coopera-
    tive Association, 166
Mayfield Swiss Cheese, 99
Mazza, Charles Sr., 125
Mazza Cheese Co., 125–26
McCoy, Pat, 138–39
McIntosh, Peter, 49, 50, 60
McKay Farm, 21, 23
McKenna, Meghan, **159**
McLoughlin, John, 19, 21, 22, 25, 27,
    28, 30
McPherson, Professor Hector, 63

Meade, A.H., 66
mechanization, 56
Menlo, WA, 99
Mexican-style cheeses, 146, 151–52
Meyers Goat Creamery, 170
Michoacano, El, 151, 170
migita cheese, 106
milk
    brought to creameries to make
        cheese, **34,** 50, 70–71, 80, 84,
        117
    clabbered, 76
    condensed, 66, 92
    industry, 66, 76, 127
    on the Oregon Trail, 37–39, 45
    raw, 113, 137, 154–56
Miller Reed Pease, 68
Millman, Mrs. Fred, 42
Minidoka Irrigation Project, 71, 88
mining, 44, 45, 70, 79, 124, 133
    See also goldsrushes
missionaries, 26–29, 164
Misterly, Rick and Lora Lea, 156, **157**
mold, 93, 98, 101, 114, 115
Monteillet Fromagerie, 170
monterey jack cheese, 117, 124, 130
Montes, Maria and Gabriel, 151
Morford, Pat, 152
Morgan, Jock, 43
Mormons, 39, 50–51, 72
Morning Star (ship), 43
Morris, Margaret, 137
Morrow, Cindy, 139
Moser, Fred, 62
Mother Earth News, 133, 134
Mountain Lodge Farm, **159,** 160, 170
Mt. Angel, OR, 116
Mt. Angel Cooperative Creamery, 85,
    86, 116
Mt. Townsend Creamery, 152, 170
mozzarella cheese, 123, 124, 125,
    137, 152
Multnomah Island
    See Sauvie Island
Musick, Mark, 134
Mystery Bay Farm, 150, 170

Mystic Lake Goat Dairy, 119

Nampa, ID, 122, **146**
Nancy the Goat, 13–14
Native Americans.
    See indigenous peoples
Neah Bay, 14
Nelson, Dennis, 142, 143
Nelson-Ricks, 146
Nelson-Ricks Cheese Factory #5, **90**
New Moon Goat Dairy, 169
Newport, OR, 140
New York, 52
Nigerian Dwarf goats, 159
Nootka Sound, 14, 30
Noris Dairy, 158, 169
North Bend, WA, 41
North Fork Cheese Factory, 62
North Pacific Cheese Factory, 49
Northwest Alternative Agriculture
    Conference, 134
North West Company, 16, 17

Oak Leaf Creamery, 157, 169
Ochoa, Francisco, **151,** 158
Ochoa's Queseria, **151,** 158, 169
oleomargarine, 112
Olympia, WA, 65
Olympia Cheese Company, 117
Olympic Peninsula, 41, 49
Olympic Ridge Creamery & Cheese
    Factory, **51**
Omnivore's Dilemma, The (Pollan), 150
100 Mile Diet, The (Smith and MacK-
    innon), 150
orchards, 19
Oregon
    artisan cheesemaking in, 149
    cheese production of, 45, 58, 103
    dairy regulations of, 155
    factory cheesemaking in, 52, 87
    state of, 55, 164
Oregon Agricultural College
    See Oregon State University
Oregon Blue cheese, 101
Oregon Cheese Guild, 150–51

Oregon Colonization Society, 35
Oregon Country, **12,** 26, 27, 30,
    35–45, 55
Oregon Dairy Goat Association, 90
Oregon Dairymen's League, 86–87
Oregon Dairy Products Commission,
    151
Oregon State Fair, 43, 62
Oregon State University, 57–58, 63,
    76, 83, 101, 139, 150, 169
Oregon Tilth, 134
Oregon Trail, **12,** 26, 35–40, 45, 50
Organic Gardening, 134
Orting, WA, 125
Our Lady of the Rock, 170
Owens, Sarah, 42
Owens-Adair, Bethenia, 43

Pacific Fur Company, 15, 16, 17
packaging, 87, 101, 102, 115, 131
    See also canned cheese; Pliofilm;
        rind; wax
Palmer, Joel, 37, 38
paneer cheese, 157
Paradise Springs Farm, 155
Paris, ID, 51
Parma, ID, 147–48
Pastega, Denny, 139
pasteurization, 113, 154
Paul Butter and Cheese Factory, **71**
Pelly, Sir John, 25, 26
Pend Oreille Cheese Co., 117
pepsin, 76
Pernot, Emil, 101
Peterson, David, 157
Pholia Farm, 155, 169
Pike Place Market, 74, 124, 138, 141,
    143
Pine Stump Farms, 170
Pittman, Anna Maria, 28
pizza, 123–24
place-based designation, 158–59
Pleasant Hill, OR, **34**
Pleasant Valley Dairy, 137, 170
Plimpton, Silas and Lydia, 40
Pliofilm, 115, 124, 130

Pocotello, ID, 39, 88
Pollan, Michael, 150
pollution, 130, 133–34
population, 16, 55, 70, 135, 165, 168
Portland, OR
    farms and dairies in and around,
        42, 47, 86, 123
    as a market for cheese, 73, 74,
        102, 140, 143
    as regional urban and industrial
        center, 55, 75
Portland Creamery, 157, 169
Port Madison Farm, 170
pot (dutch) cheese, 38, 43, 165
presses, 44, 47, 56
Pritchard, James, 39
prizes, 46, 139, 148, 152
processed cheese, 87–88, 89
provolone cheese, 123, 124
Puget Sound, 20, 65, 68, 133
Puget Sound Agricultural Company
    (PSAC), 20, 23, 31, 153, 164

Quail Croft Farm, 170
Quail Run Creamery, 169
quality control, 56, 60, 64, 84, 87,
    101, 113, 127
quark cheese, 157
Queseria Bendita, 151, 170
queso fresco, 151
queso oaxaca, 151
Quillisascut Farmstead Cheese, 156,
    157, 170
Quimby Farm, 42

railroads, 49, 55, 56, 57, 59, 61, 68,
    80, 165
Rainier, OR, 40
raw milk, 113, 137, 154–56
Red River Settlement, 18
Red Rock Cheese Company, 89
Reed, Briggs F., 68
refrigeration, 45, 101, 115
regulation, 113–14, 127, 130,
    154–55, 167
rennet, 22, 28, 56, 76, 157

retail stores, 122, 138, 141–42,
    143–44, 152, 168
Reuter, Frederick, **66,** 67
Reynolds, Dr. Frank, 165
Rich, Charles, 51
rind, 61, 101, 114–15, 123
ripening, 91, 93
Rivers Edge Chevre, 152, 169
River Valley Ranch, 171
Roberts, Jeffrey, 149, 152, 154
Rocky Mountains, 17, 26, 50, 121,
    140, 148, 159
Rogue Creamery, **11,** 101, 142, 152,
    158, 169
Rogue River Valley Creamery,
    99–101, 116
Rollingstone Chevre, 147–48, 171
Roquefort cheese, 91, 93, 96
Rosecrest Farm, 171
Russell, Kendall, 148
Russian-American Company, 15, 20,
    24, 31, 163–64
Russian exploration of Pacific North-
    west, 13, 14, 15, 20, 24, 163–64

Saanen goats, 78
Sacred Heart mission, 29
Safeway Corporation, 116, 118
Sahli, Adolf, **54**
St. John Monastery Farm, 171
St. Paul, OR, 27
Salem, OR, 46, 140
sales
    See marketing and sales
salting, **11,** 91
Samish Bay Cheese Co., 171
Sandwich Islands, 19, 20
Sartori, 146
Saunders, Clarence, 75
Saunders Cheese Shop, 144
Sauvie Island, 21, 23, 24
Scappoose, OR, 21
Schooler, Luan, 143
Seattle, WA
    as a market for cheese, **69,** 73, 74,
        140, 141–42, 143

    as regional center, 55, 102
Seattle Cheese Co., 124
Sharp, J. P., 69
sheep ranching, 148
sheep's milk cheese, 62, 148, 149
shops, 122, 138, 141–42, 143–44,
    152, 168
Shorthorn cattle, 56
Sikkens, Peter, 139
Silver Springs Creamery, 171
Silverton, OR, 46
Simeonof Island, 165
Simpson, George, 17–19, 20, 23, 28,
    29
Simpson, Theresa, 142, 143
Sitka, AK, 20
Skagit Valley, 124, 159
Sluder, Laura, 148, 158
Smith, Alisa, 150
Smith, Jasper, 59
Snake River, 70, 71, 133
Sonoma Valley Cheese Factory, **100**
Sorrento Cheese Factory, **146**
Sorrento-Lactalis, 144, 146
Spaulding, Henry and Eliza, 27, 69
Spencer, Homer, 97–98
spoiled cheese, 101, 114, 115
Spokane, WA, 19, 28, 67, 73
Spokane Indians, 28
standardization, 61, 111–14, 154
Stanfield Cheese Factory, 54
Star of Oregon (ship), 30
Star Ranch, 94, 95
Star Valley Swiss Cheese Co., 121,
    123
Stayton, OR, 48
Steamboat Island Goat Farm, 171
Stephens, Sydney, 44
Steve's Cheese, 143, 144
Stiles, C.T., 47
Stilton cheese, 24, 96
Sumner Creamery, 62
Sunnydale Goat Farm, 79
Sunny Pine Farm, 171
Superior Cheese Co., 118

sustainable farming, 134, 135, 150, 160

"sweet cheddar," 60

swiss cheese, 51, 97, 120–23, 148

Swiss Village Cheese Co., 122

Tacoma, WA, 20, 74

Tenas Illihee Island, 105, **106**

territorial claims to PacificWNorthwest, 14–16, 18, 26, 29, 30, 31, 35

terroir, 159

Teton Valley Creamery, 148, 171

The Dalles, OR, 27, 39, 44

thistle rennet, 157

Threemile Canyon Farms, 147

Tieton Farm and Creamery, 171

Tigard, OR, 89

Tillamook, OR, 40, 43, 50, 58–59, 83, 126–28, 139, 147

Tillamook cheese, 63, 73–74, 76, 80, 109, **114,** 115, 167

Tillamook Cheese Association, 109

Tillamook Cheese Factory, **108,** 109, 111, 126, 139

Tillamook City Creamery, 59

Tillamook County Creamery Association (TCCA)
    formation of, 60–61
    production of, 103, 110, 147
    relations with other cooperatives, 63–64, 126–28, 130

Tillamook Creamery, **60,** 109

Tillamook Dairy and Cheese Association (TCDA), 126–28, 130

Timmermeister, Kurt, 156

Toggenburg goats, 78, 79

Toledo, WA, 20, 153

Toledo Cheese Days, 153

Townsend, Thaddeus (T. S.), 50

Train, George, 137

transhumance, 73

Trout Lake Valley, 96

Tshimakin mission, 28

tuberculosis, 76–78, 113

Tumalo Farms Cheese Co., **143,** 169

Turnbull, Amy, 149

Twin Falls, ID, 44, 72, 110

Twin Oaks Creamery, 171

unemployment, 83

United Dairymen's Association of Washington, 85, 111

United States
    back to the land tradition in, 133
    involvement in WWI and WWII, 75, 103
    population and economic growth, 55–56
    territorial claims to PacificWNorthwest, 16, 30
    westward expansion of, 26, 30, 35–36, 164

United States Department of Agriculture (USDA), 101, 104, 120, 134, 152, 165

University of Idaho, 57, 67

Upper Snake River Dairymen's Association, 110

Valdez Dairy, 165

Vancouver, BC, 19

Vancouver, WA, 75, 102

Vancouver Island, 14, 30, 31

vats, 47, 56, 153

Vaughn, Warren, 43

Vella, Gaetano "Tom," 100–101

Vella, Ignazio Ig," 100

Victoria, BC, 42

Viewfield Farm, 31

Viviani, Celso, 100

Volbeda, Rod, 140, 150

Waiilatpu mission, 27, 28, 69

Walker, Mary, 28

Walker, Tom, cheese factory of, **34**

Wallace, Henry, 83

Walla Walla, WA, 21, 22, 27

Wapato Island
    See Sauvie Island

War of 1812, 16

Warrenton, OR, 42

Wascopam mission, 27, 28, 39

Washington, 45, 55, 66, 85, 98, 138, 149, 155, 164

Washington Mountain Cheese Factory, 106

Washington State University, 57, **82,** 102, 143, 150, 156

Washington State University Creamery, 171

Washington Territory, 31

water, 40, 70, 72

wax packaging, 115

Werlin, Laura, 157–58

West, Harry, 89

West, Helen, 89

West, Josiah, 47

Western Dairy Products Company, 68

Western Goat Products, 92

Whatcom County Dairymen's Association, 111

Wheeler, Ernest, **34**

whey, 91, 106, 145

Wheyward Goat Cheese Co., 148, 149, 171

Whipple, Simon, 68

Whiskey Hill Farm, 171

White Clover butter, 50

Whitman, Marcus and Narcissa, 22–23, 26–27, 28, 69

Whole Earth Catalog, 134

Wild Goose Farm, 169

Wild Harvest Creamery, 171

Wilkes, Charles, 23, 24

Willamette River, 18, 55, 133

Willamette Valley, 27, 30, 159

Willamette Valley Cattle Company, 30

Willamette Valley Cheese Co., 140, 150, 169

Willapa Hills Farmstead Cheese, 149, 171

Wilson, Tim, 143

Wimberley, Susanne, 149

Windsong Farms, 167

wine industry, 138, 140, 159

Winship, Nathan, 15, 164

Wisconsin, 60, 88, 117, 121, 123, 144, 157

women, 24, 25, 28, 37, 41, 100

Woodland Co-Operative Cheese Factory, 49

Woodland Dairy Association, 49

Wood n' Goat Garden, 171

Woodrich, Adolph, 129

World War I, 75–76, 83, 88, 107

World War II, 100, 102–4, 107, 115, 123

Yakima River, 70

Yakima Valley Cheese Company, 139

Yarmuth Farms, 171

Yoncalla, OR, 44

Young, Ewing, 30